STUDENT'S
SOLUTIONS MANUAL

ELEMENTARY
ALGEBRA
4TH EDITION

STUDENT'S
SOLUTIONS MANUAL

E L E M E N T A R Y
A L G E B R A
4 TH EDITION

DANIEL L. AUVIL / CHARLES POLUGA
KENT STATE UNIVERSITY

ADDISON-WESLEY PUBLISHING COMPANY
Reading, Massachusetts • Menlo Park, California • New York
Don Mills, Ontario • Wokingham, England • Amsterdam • Bonn
Sydney • Singapore • Tokyo • Madrid • San Juan • Milan •Paris

Reprinted with corrections, October 1992.

Reproduced by Addison-Wesley from camera-ready copy supplied by the author.

ISBN 0-201-14987-7

2 3 4 5 6 7 8 9 10-AL-95949392

TABLE OF CONTENTS

STUDENT'S
SOLUTIONS MANUAL

ELEMENTARY
ALGEBRA
4TH EDITION

CHAPTER 1

REAL NUMBERS AND THEIR PROPERTIES

Problem Set 1.1, pp. 10-12

1. Factors of 6:1, 2, 3, 6

   Factors of 8:1, 2, 4, 8

   GCF = 2

3. Factors of 4:1, 2, 4

   Factors of 20:1, 2, 4, 5, 10, 20

   GCF = 4

5. Factors of 12:1, 2, 3, 4, 6, 12

   Factors of 18:1, 2, 3, 6, 9, 18

   GCF = 6

7. Factors of 2:1, 2

   Factors of 21:1, 3, 7, 21

   GCF = 1

9. $\dfrac{2}{4} = \dfrac{2 \div 2}{4 \div 2} = \dfrac{1}{2}$

11. $\dfrac{8}{12} = \dfrac{8 \div 4}{12 \div 4} = \dfrac{2}{3}$

13. $\dfrac{15}{20} = \dfrac{15 \div 5}{20 \div 5} = \dfrac{3}{4}$

15. $\dfrac{30}{54} = \dfrac{30 \div 6}{54 \div 6} = \dfrac{5}{9}$

17. $\dfrac{60}{48} = \dfrac{60 \div 12}{48 \div 12} = \dfrac{5}{4}$

19. $\dfrac{240}{800} = \dfrac{240 \div 80}{800 \div 80} = \dfrac{3}{10}$

21. $\dfrac{3}{8} \cdot \dfrac{4}{9} = \dfrac{\overset{1}{\cancel{3}}}{\underset{2}{\cancel{8}}} \cdot \dfrac{\overset{1}{\cancel{4}}}{\underset{3}{\cancel{9}}} = \dfrac{1}{6}$

23. $\dfrac{1}{2} \cdot \dfrac{1}{2} = \dfrac{1 \cdot 1}{2 \cdot 2} = \dfrac{1}{4}$

25. $4 \cdot \dfrac{3}{40} = \dfrac{\overset{1}{\cancel{4}}}{1} \cdot \dfrac{3}{\underset{10}{\cancel{40}}} = \dfrac{3}{10}$

27. $\dfrac{2}{7} \div \dfrac{3}{5} = \dfrac{2}{7} \cdot \dfrac{5}{3} = \dfrac{10}{21}$

29. $\dfrac{1}{2} \div 2 = \dfrac{1}{2} \cdot \dfrac{1}{2} = \dfrac{1}{4}$

31. $\dfrac{4}{9} \div \dfrac{3}{4} = \dfrac{4}{9} \cdot \dfrac{4}{3} = \dfrac{16}{27}$

33. $\dfrac{3}{7} \cdot 21 = \dfrac{3}{\underset{1}{\cancel{7}}} \cdot \dfrac{\overset{3}{\cancel{21}}}{1} = \dfrac{9}{1} = 9$

35. $\dfrac{9}{2} \div \dfrac{3}{2} = \dfrac{\overset{3}{\cancel{9}}}{\underset{1}{\cancel{2}}} \cdot \dfrac{\overset{1}{\cancel{2}}}{\underset{1}{\cancel{3}}} = \dfrac{3}{1} = 3$

37. $\dfrac{1}{4} + \dfrac{1}{4} = \dfrac{1+1}{4} = \dfrac{2}{4} = \dfrac{1}{2}$

39. $\dfrac{5}{8} + \dfrac{1}{8} = \dfrac{5+1}{8} = \dfrac{6}{8} = \dfrac{3}{4}$

41. $\dfrac{5}{7} - \dfrac{3}{7} = \dfrac{5-3}{7} = \dfrac{2}{7}$

43. $\dfrac{13}{20} - \dfrac{7}{20} = \dfrac{13-7}{20} = \dfrac{6}{20} = \dfrac{3}{10}$

45. $\dfrac{2}{3} + \dfrac{1}{9} = \dfrac{6}{9} + \dfrac{1}{9} = \dfrac{7}{9}$

47. $\dfrac{1}{2} + \dfrac{1}{5} = \dfrac{5}{10} + \dfrac{2}{10} = \dfrac{7}{10}$

49. $\dfrac{8}{15} - \dfrac{2}{5} = \dfrac{8}{15} - \dfrac{6}{15} = \dfrac{2}{15}$

51. $\dfrac{1}{3} - \dfrac{1}{4} = \dfrac{4}{12} - \dfrac{3}{12} = \dfrac{1}{12}$

53. $\dfrac{1}{6} + \dfrac{5}{9} = \dfrac{3}{18} + \dfrac{10}{18} = \dfrac{13}{18}$

55. $\dfrac{5}{6} - \dfrac{3}{4} = \dfrac{10}{12} - \dfrac{9}{12} = \dfrac{1}{12}$

57. $6 - \dfrac{3}{7} = \dfrac{42}{7} - \dfrac{3}{7} = \dfrac{39}{7}$

59. $\dfrac{2}{5} + \dfrac{1}{4} + \dfrac{3}{10} = \dfrac{8}{20} + \dfrac{5}{20} + \dfrac{6}{20} = \dfrac{19}{20}$

61. $3\dfrac{1}{2} \cdot 2\dfrac{1}{5} = \dfrac{7}{2} \cdot \dfrac{11}{5} = \dfrac{77}{10}$

63. $2\dfrac{2}{5} \div 1\dfrac{1}{3} = \dfrac{12}{5} \div \dfrac{4}{3} = \dfrac{\overset{3}{\cancel{12}}}{5} \cdot \dfrac{3}{\underset{1}{\cancel{4}}} = \dfrac{9}{5}$

65. $4\dfrac{1}{2} + 2\dfrac{1}{5} = \dfrac{9}{2} + \dfrac{11}{5} = \dfrac{45}{10} + \dfrac{22}{10} = \dfrac{67}{10}$

67. $8 - 3\dfrac{2}{3} = \dfrac{24}{3} - \dfrac{11}{3} = \dfrac{13}{3}$

69. $\dfrac{22{,}500}{25{,}000} = \dfrac{9}{10}$

71. $12 \cdot 8\dfrac{1}{4} = \dfrac{\overset{3}{\cancel{12}}}{1} \cdot \dfrac{33}{\underset{1}{\cancel{4}}} = \$99$

73. $10\dfrac{1}{8} - 7\dfrac{1}{2} = \dfrac{81}{8} - \dfrac{15}{2} = \dfrac{81}{8} - \dfrac{60}{8} = \dfrac{21}{8} = \$2\dfrac{5}{8}$

75. $24 \div 2\dfrac{2}{3} = \dfrac{24}{1} \div \dfrac{8}{3} = \dfrac{\overset{3}{\cancel{24}}}{1} \cdot \dfrac{3}{\underset{1}{\cancel{8}}} = 9 \text{ glasses}$

77. Total distance $= 5\dfrac{1}{2} + 3\dfrac{2}{5} + 7\dfrac{3}{5} + 6 = \dfrac{11}{2} + \dfrac{17}{5} + \dfrac{38}{5} + \dfrac{6}{1}$

$$= \dfrac{55}{10} + \dfrac{34}{10} + \dfrac{76}{10} + \dfrac{60}{10}$$

$$= \dfrac{225}{10} = \dfrac{45}{2}$$

Average $= \dfrac{45}{2} \div 4 = \dfrac{45}{2} \cdot \dfrac{1}{4} = \dfrac{45}{8} = 5\dfrac{5}{8} \text{ mi}$

79. $10 - (2\frac{3}{4} + 3\frac{1}{2}) = 10 - (\frac{11}{4} + \frac{7}{2}) = 10 - (\frac{11}{4} + \frac{14}{4}) = \frac{40}{4} - \frac{25}{4} = \frac{15}{4} = 3\frac{3}{4}$ ton

81. $\frac{12}{16} = \frac{3}{4}$ and $\frac{3}{4}$ of $\frac{2}{3} = \frac{\overset{1}{\cancel{3}}}{\underset{2}{\cancel{4}}} \cdot \frac{\overset{1}{\cancel{2}}}{\underset{1}{\cancel{3}}} = \frac{1}{2}$ cup

Problem Set 1.2, pp. 16-18

1. +18° means 18° above zero.

    -4° means 4° below zero.

3. $+1\frac{3}{8}$ means a rise of $1\frac{3}{8}$.

    $-3\frac{1}{8}$ means a fall of $3\frac{1}{8}$.

5. +6 yd means a gain of 6 yd.

    -3 yd means a loss of 3 yd.

7. False. -78 is not an element
    of the set $\{0, 1, 2, 3, \ldots\}$.

9. True. 0 is an element of the
    set $\{\ldots, -2, -1, 0, 1, 2, \ldots\}$.

11. False. 0 is neither positive
    nor negative.

13. True. $\frac{2}{3}$ is a quotient of two
    integers.

15. False. $6\frac{1}{4} = \frac{25}{4}$, which is a
    quotient of integers and
    therefore a rational number.

17. True. For example, $-5 = \frac{-5}{1}$.

19. True. Every element of
    $\{0, 1, 2, \ldots\}$ is also an
    element of
    $\{\ldots, -2, -1, 0, 1, 2, \ldots\}$.

21. True. For example, $\frac{3}{4}$.

23. True

25. True. For example, $-\frac{3}{4}$.

27. True

29. -1, -4, and -10

31. $\sqrt{2}$, $\sqrt{3}$, and $\sqrt{5}$

33. -5, 0, and $\frac{1}{2}$

35. (a) 18, 0, $\sqrt{9}$

     (b) 18, -10, 0, $\sqrt{9}$

     (c) $\frac{5}{7}$, 18, 0.35, -10, 0, $\sqrt{9}$, $-3\frac{1}{6}$

     (d) $\sqrt{6}$, $-\sqrt{2}$

     (e) All of them

37. -2

39. 63

41. $\frac{5}{6}$

43. -0.125

45. $9\frac{1}{4}$

47. 0

49. $-\sqrt{3}$

51. $\sqrt{15}$

53. -(-5) = 5

55. $-(-4\frac{2}{5}) = 4\frac{2}{5}$

57. -(-8.6) = 8.6

59. -(-(-17)) = -(17) = -17

61.

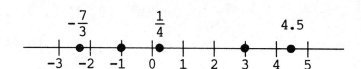

63. A positive number. For example, if x = -3, then -x = -(-3) = 3.

65. ┌─────┐   ┌─────┐
     │Clear│   1   │ +/- │   -1
     └─────┘   └─────┘

     The opposite of 1 is -1.

67. ┌─────┐   ┌─────┐
     │Clear│   0.375   │ +/- │   -0.375
     └─────┘   └─────┘

     The opposite of 0.375 is -0.375.

69. ┌─────┐   ┌─────┐
     │Clear│   3   │ √x │   1.7320508
     └─────┘   └─────┘

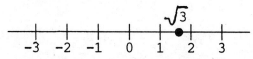

71. ┌─────┐   ┌─────┐   ┌─────┐
     │Clear│   7   │ √x │   │ +/- │   -2.6457513
     └─────┘   └─────┘   └─────┘

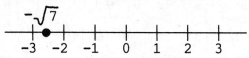

73. ┌─────┐   ┌─────┐   ┌─────┐
     │Clear│   4   │ +/- │   │ √x │   Error
     └─────┘   └─────┘   └─────┘

     $\sqrt{-4}$ is not a real number.

Problem Set 1.3, pp. 22-24

1. $5 > 3$

3. $7.07 < 7.1$

5. $-7 < 0$

7. $-4 < -2$

9. $0.3 > -0.6$

11. $-1\frac{3}{5} < -1\frac{2}{5}$

13. $\frac{3}{4} > \frac{5}{7}$ since $\frac{3}{4} = \frac{21}{28}$ and $\frac{5}{7} = \frac{20}{28}$.

15. $x = -17$

17. $z < 8$

19. $r \geq -6$

21. $p \leq 7$

23. $x > 0$

25. $y \geq 0$

27. $a \geq 21$ yr

29. $w \leq 160$ lb

31. $4 < x < 9$

33. $12 \geq y > -1$

35. $-3 < z < 6$

37. $3 \leq p < 10$

39. $-8° \leq t \leq 19°$

41. If you drive the minimum speed for 3 hr, then $d = 35 \cdot 3 = 105$ mi.

    If you drive the maximum speed for 3 hr, then $d = 55 \cdot 3 = 165$ mi.

    Therefore $105 \text{ mi} \leq d \leq 165 \text{ mi}$.

43. If the total cost is \$450, then $C = \frac{450}{40} = \$11.25$. If the total cost
    is \$600, then $C = \frac{600}{40} = \$15$. Therefore $\$11.25 \leq C \leq \$15$.

45. $|6| = 6$

47. $|-6| = 6$

49. $|-1| = 1$

51. $\left|-\frac{3}{4}\right| = \frac{3}{4}$

53. $|7.2| = 7.2$

55. $|-\sqrt{2}| = \sqrt{2}$

57. $-|10| = -10$

59. $-|-10| = -10$

61. $|8 - 3| = |5| = 5$

63. $-|8 - 3| = -|5| = -5$

65. $|-9| - |2| = 9 - 2 = 7$

67. $|-6| - |-4| = 6 - 4 = 2$

69. $|8.1| > |5.96|$

71. $|-25| > |0|$

73. $|1| = |-1|$

75. $|3| < |-9|$

77. $|-14| < |18|$

79. $|-\frac{1}{2}| < |-\frac{3}{4}|$

81. 8 and -8

83. 4 and -8

85. False. For example, if a = -3, then $|a| = |-3| = 3 \neq a$.

87. False. For example, if a = -5 and b = 2, then $-5 < 2$ but $|-5| \not< |2|$.

Problem Set 1.4, pp. 27-28

1. $4 + 2 = 6$

3. $-2 + (-5) = -7$

5. $-6 + (-9) = -15$

7. $-3 + 7 = 4$

9. $1 + (-5) = -4$

11. $-9 + 5 = -4$

13. $10 + (-3) = 7$

15. $-8 + (-6) = -14$

17. $18 + (-5) = 13$

19. $2 + (-17) = -15$

21. $-14 + 13 = -1$

23. $11 + (-7) = 4$

25. $\quad 2 + 6 = 8$

27. $\quad\quad 24.1 + 9.2 = 33.3$

$\quad -2 + (-6) = -8$

$\quad -24.1 + (-9.2) = -33.3$

$\quad 2 + (-6) = -4$

$\quad 24.1 + (-9.2) = 14.9$

$\quad -2 + 6 = 4$

$\quad -24.1 + 9.2 = -14.9$

29. $1 + (-1) = 0$

31. $-8 + 0 = -8$

33. $-156 + 156 = 0$

35. $0 + (-93) = -93$

37. $-\frac{2}{7} + (-\frac{3}{7}) = -\frac{5}{7}$

39. $-\frac{1}{3} + \frac{5}{6} = -\frac{2}{6} + \frac{5}{6} = \frac{3}{6} = \frac{1}{2}$

41. $-\frac{4}{5} + \frac{4}{5} = 0$

43. $\frac{2}{3} + (-\frac{3}{4}) = \frac{8}{12} + (-\frac{9}{12}) = -\frac{1}{12}$

45. 6.8 + (-1.5) = 5.3

47. -8.3 + (-5.94) = -14.24

49. 4.7 + (-4.7) = 0

51. -4.1 + 3.62 = -0.48

53. 1 + 5 + (-4) + (-2) = 6 + (-6) = 0

55. 4 + (-6) + 8 + (-10) = 12 + (-16) = -4

57. 18 + (-3) + (-3) + (-3) = 18 + (-9) = 9

59. -27 + 17 + 10 + 31 = -27 + 58 = 31

61. -10, since -10 + 6 = -4

63. 8, since 8 + (-5) = 3

65. -8 + 21 = 13°

67. 2 + 0 + 1 + (-2) + (-3) + (-2) + 0 + (-1) + 3 = 6 + (-8) = -2

69. 3.67 + (-4.19) + 1.12 + (-0.92) + 6.43 = 11.22 + (-5.11) = 6.11 up

71. | Clear | 94 | +/- | | + | 38 | = | -56

-94 + 38 = -56

73. | Clear | 882 | +/- | | + | 579 | +/- | | = | -1461

-882 + (-579) = -1461

Problem Set 1.5, pp. 31-32

1. 3 - 7 = 3 + (-7) = -4

3. -5 - 4 = -5 + (-4) = -9

5. 9 - (-6) = 9 + 6 = 15

7. 4 - 8 = 4 + (-8) = -4

9. -3 - 3 = -3 + (-3) = -6

11. 6 - (-1) = 6 + 1 = 7

13. -12 - 3 = -12 + (-3) = -15

15. 10 - 15 = 10 + (-15) = -5

17. 11 - (-12) = 11 + 12 = 23

19. -8 - (-2) = -8 + 2 = -6

21. -9 - 8 = -9 + (-8) = -17

23. 16 - 0 = 16 + 0 = 16

25. -1 - (-7) = -1 + 7 = 6

27. -15 - (-6) = -15 + 6 = -9

29. 0 - 5 = 0 + (-5) = -5

31. 0 - (-10) = 0 + 10 = 10

33. $93 - 49 = 93 + (-49) = 44$

35. $70 - (-70) = 70 + 70 = 140$

37. $17 - 189 = 17 + (-189) = -172$

39. $110 - 110 = 110 + (-110) = 0$

41. $-52 - 0 = -52 + 0 = -52$

43. $-200 - (-200) = -200 + 200 = 0$

45. $432 - 167 = 432 + (-167) = 265$

$-432 - 167 = -432 + (-167) = -599$

$432 - (-167) = 432 + 167 = 599$

$-432 - (-167) = -432 + 167 = -265$

47. $5\frac{1}{2} - 3\frac{3}{4} = 5\frac{1}{2} + (-3\frac{3}{4}) = \frac{11}{2} + (-\frac{15}{4}) = \frac{22}{4} + (-\frac{15}{4}) = \frac{7}{4} = 1\frac{3}{4}$

$-5\frac{1}{2} - 3\frac{3}{4} = -5\frac{1}{2} + (-3\frac{3}{4}) = -\frac{11}{2} + (-\frac{15}{4}) = -\frac{22}{4} + (-\frac{15}{4}) = -\frac{37}{4} = -9\frac{1}{4}$

$5\frac{1}{2} - (-3\frac{3}{4}) = 5\frac{1}{2} + 3\frac{3}{4} = \frac{11}{2} + \frac{15}{4} = \frac{22}{4} + \frac{15}{4} = \frac{37}{4} = 9\frac{1}{4}$

$-5\frac{1}{2} - (-3\frac{3}{4}) = -5\frac{1}{2} + 3\frac{3}{4} = -\frac{11}{2} + \frac{15}{4} = -\frac{22}{4} + \frac{15}{4} = -\frac{7}{4} = -1\frac{3}{4}$

49. $\frac{1}{3} - \frac{1}{2} = \frac{1}{3} + (-\frac{1}{2}) = \frac{2}{6} + (-\frac{3}{6}) = -\frac{1}{6}$

51. $\frac{1}{3} - \frac{5}{9} = \frac{1}{3} + (-\frac{5}{9}) = \frac{3}{9} + (-\frac{5}{9}) = -\frac{2}{9}$

53. $-\frac{3}{4} - (-\frac{5}{6}) = -\frac{3}{4} + \frac{5}{6} = -\frac{9}{12} + \frac{10}{12} = \frac{1}{12}$

55. $1\frac{3}{5} - 6 = 1\frac{3}{5} + (-6) = \frac{8}{5} + (-\frac{30}{5}) = -\frac{22}{5} = -4\frac{2}{5}$

57. $-5.7 - 3.1 = -5.7 + (-3.1) = -8.8$

59. $4.99 - 8.3 = 4.99 + (-8.3) = -3.31$

61. $6.84 - (-5.2) = 6.84 + 5.2 = 12.04$

63. $-1.9 - (-2.3) = -1.9 + 2.3 = 0.4$

65.   -2 + 7 = 5 suggests that 5 - (-2) = 7 and 5 - 7 = -2.

67.   -1 + (-3) = -4 suggests that -4 - (-1) = -3 and -4 - (-3) = -1.

69.   2 - 8 = 2 + (-8) = -6

71.   6 - (-13) = 6 + 13 = 19

73.   -23 - 14 = -23 + (-14) = -37

75.   -2 - 9 = -2 + (-9) = -11°

77.   14,490 - (-280) = 14,490 + 280 = 14,770 ft

79.   -8.36 - 7.67 = -8.36 + (-7.67) = $-16.03

81. | Clear | 63 | - | 91 | = | -28

      63 - 91 = -28

83. | Clear | 559 | +/- | | - | 712 | +/- | | = | 153

      -559 - (-712) = 153

Problem Set 1.6, pp. 36-37

1.   3 · 10 = 30

      2 · 10 = 20

      1 · 10 = 10

      0 · 10 = 0

      -1 · 10 = -10

      -2 · 10 = -20

3.   (-8)2 = -16   5.   5(-4) = -20

7.   6 · 7 = 42   9.   -2(-3) = 6

11. -1(-1) = 1   13. 16(-1) = -16

15. -23 · 0 = 0   17. 6(-6) = -36

19. (-9)1 = -9   21. (-12)(-12) = 144

23. $(-15)(-4) = 60$

25. $0(-6) = 0$

27. $6(8) = 48$

   $-6(-8) = 48$

   $6(-8) = -48$

   $(-6)8 = -48$

29. $168(4.2) = 705.6$

   $-168(-4.2) = 705.6$

   $168(-4.2) = -705.6$

   $(-168)4.2 = -705.6$

31. $\left(-\dfrac{3}{4}\right)\left(-\dfrac{5}{7}\right) = \dfrac{15}{28}$

33. $\left(-\dfrac{2}{9}\right)\dfrac{5}{2} = -\dfrac{10}{18} = -\dfrac{5}{9}$

35. $10\left(-\dfrac{1}{2}\right) = -\dfrac{10}{2} = -5$

37. $3\left(-\dfrac{1}{3}\right) = -\dfrac{3}{3} = -1$

39. $\left(-\dfrac{1}{6}\right)(-6) = \dfrac{6}{6} = 1$

41. $(-1.1)(-1.1) = 1.21$

43. $13.9(-6.7) = -93.13$

45. $(-0.08)(-0.02) = 0.0016$

47. $(-7.35)4.3 = -31.605$

49. $(-2)(-5)(-7) = (10)(-7) = -70$

51. $(-4)(4)(-4) = (-16)(-4) = 64$

53. $(-1)(-4)(-2)(-3) = (4)(6) = 24$

55. $10(-6)(-2)(-5) = (-60)10 = -600$

57. $9(-2)(0)(7)(-8) = 0$

59. $\dfrac{1}{5}$

61. -6

63. $-\dfrac{3}{2}$

65. 1

67. Since $-0.1 = -\dfrac{1}{10}$, the reciprocal is -10.

69. Since $2\dfrac{3}{7} = \dfrac{17}{7}$, the reciprocal is $\dfrac{7}{17}$.

71. $3(-4) = -12$ yd

73. $6(-2) = -12$ lb

75. $5(-500) = \$-2500$

77. | Clear | 19 | x | 14 | +/- | = | -266

   $19(-14) = -266$

79. | Clear | 182 | +/- | | x | 653 | +/- | | = | 118,846

     (-182)(-653) = 118,846

81. | Clear | 2 | • | 5 | +/- | | $\frac{1}{x}$ | -0.4

     $\frac{1}{-2.5}$ = -0.4

Problem Set 1.7, pp. 41-43

1.   15 ÷ 5 = 3

      10 ÷ 5 = 2

       5 ÷ 5 = 1

       0 ÷ 5 = 0

      -5 ÷ 5 = -1

     -10 ÷ 5 = -2

3. $\frac{10}{2}$ = 5

5. $\frac{12}{-3}$ = -4

7. $\frac{-15}{-5}$ = 3

9. $\frac{-20}{5}$ = -4

11. $\frac{34}{-17}$ = -2

13. $\frac{-27}{-9}$ = 3

15. $\frac{-25}{-5}$ = 5

17. $\frac{-6}{1}$ = -6

19. $\frac{-18}{18}$ = -1

21. $\frac{-280}{7}$ = -40

23. $\frac{1}{-1}$ = -1

25. $\frac{-56}{-1}$ = 56

27. $\frac{180}{-20}$ = -9

29. $\frac{-23}{-23}$ = 1

31. $\dfrac{16}{2} = 8$

$\dfrac{-16}{-2} = 8$

$\dfrac{-16}{2} = -8$

$\dfrac{16}{-2} = -8$

33. $6 \div \dfrac{2}{3} = \dfrac{6}{1} \cdot \dfrac{3}{2} = \dfrac{18}{2} = 9$

$-6 \div \left(-\dfrac{2}{3}\right) = -\dfrac{6}{1}\left(-\dfrac{3}{2}\right) = \dfrac{18}{2} = 9$

$-6 \div \dfrac{2}{3} = -\dfrac{6}{1} \cdot \dfrac{3}{2} = -\dfrac{18}{2} = -9$

$6 \div \left(-\dfrac{2}{3}\right) = \dfrac{6}{1}\left(-\dfrac{3}{2}\right) = -\dfrac{18}{2} = -9$

35. $\dfrac{-2}{-5} = \dfrac{2}{5}$

37. $\dfrac{-12}{30} = -\dfrac{12}{30} = -\dfrac{2}{5}$

39. $-\dfrac{-3}{8} = \dfrac{-3}{-8} = \dfrac{3}{8}$

41. $\dfrac{10}{-25} = -\dfrac{10}{25} = -\dfrac{2}{5}$

43. $-\dfrac{4}{-16} = \dfrac{-4}{-16} = \dfrac{4}{16} = \dfrac{1}{4}$

45. $-\dfrac{-21}{-28} = -\dfrac{21}{28} = -\dfrac{3}{4}$

47. $\dfrac{-3}{8} \div \dfrac{2}{5} = -\dfrac{3}{8} \cdot \dfrac{5}{2} = -\dfrac{15}{16}$

49. $\dfrac{1}{2} \div (-2) = \dfrac{1}{2}\left(-\dfrac{1}{2}\right) = -\dfrac{1}{4}$

51. $\dfrac{-15}{4} \div \dfrac{5}{-6} = -\dfrac{\overset{3}{\cancel{15}}}{\underset{2}{\cancel{4}}}\left(-\dfrac{\overset{3}{\cancel{6}}}{\underset{1}{\cancel{5}}}\right) = \dfrac{9}{2}$

53. $-18 \div \left(\dfrac{9}{-10}\right) = -\dfrac{\overset{2}{\cancel{18}}}{1}\left(-\dfrac{10}{\underset{1}{\cancel{9}}}\right) = 20$

55. $-21.44 \div 6.7 = -3.2$

57. $-10.8 \div (-0.72) = 15$

59. $328.02 \div (-21.3) = -15.4$

61. $\dfrac{0}{5} = 0$

63. $\dfrac{-4}{0}$ is undefined.

65. $0 \div 13 = 0$

67. $-1 \div 0$ is undefined.

69. $4(-6) = -24$ suggests that $-24 \div 4 = -6$ and $-24 \div (-6) = 4$.

71. $(-5)(-8) = 40$ suggests that $40 \div (-5) = -8$ and $40 \div (-8) = -5$.

73. $\dfrac{6}{-12} = -\dfrac{6}{12} = -\dfrac{1}{2}$

75. $\dfrac{-10}{0}$ is undefined.

77. $\dfrac{-21}{7} = -3°$

79.   Ave = $\dfrac{-3 + 2 + 1 + 0 + (-9) + (-3)}{6} = \dfrac{-12}{6} = -2$ yd

81. $\dfrac{0}{4} = 0$ since $4 \cdot 0 = 0$.

83. $\boxed{\text{Clear}}$ 0 $\boxed{\div}$ 4 $\boxed{+/-}$ $\boxed{=}$ 0

$0 \div (-4) = 0$

85. $\boxed{\text{Clear}}$ 271.86 $\boxed{+/-}$ $\boxed{\div}$ 19.7 $\boxed{=}$ -13.8

$-271.86 \div 19.7 = -13.8$

Problem Set 1.8, pp. 46-48

1. $8^2 = 8 \cdot 8 = 64$

3. $10^4 = 10 \cdot 10 \cdot 10 \cdot 10 = 10,000$

5. $(-5)^2 = (-5)(-5) = 25$

7. $-5^2 = -(5 \cdot 5) = -25$

9. $(-5)^3 = (-5)(-5)(-5) = -125$

11. $-5^3 = -(5 \cdot 5 \cdot 5) = -125$

13. $\left(\dfrac{3}{4}\right)^4 = \left(\dfrac{3}{4}\right)\left(\dfrac{3}{4}\right)\left(\dfrac{3}{4}\right)\left(\dfrac{3}{4}\right) = \dfrac{81}{256}$

15. $\left(-\dfrac{1}{2}\right)^2 = \left(-\dfrac{1}{2}\right)\left(-\dfrac{1}{2}\right) = \dfrac{1}{4}$

17. $-2^4 = -(2 \cdot 2 \cdot 2 \cdot 2) = -16$

19. $(-1)^6 = (-1)(-1)(-1)(-1)(-1)(-1) = 1$

21. $2 \cdot 5^2 = 2 \cdot 25 = 50$

23. $12 - 3^2 = 12 - 9 = 3$

25. $5 + 2 \cdot 3 = 5 + 6 = 11$

27. $18 - 5 + 3 = 13 + 3 = 16$

29. $16 \div 8 \cdot 2 = 2 \cdot 2 = 4$

31. $3 + 12 \div 3 + 3 = 3 + 4 + 3 = 10$

33. $6 - 3 + 4^2 - 2 = 6 - 3 + 16 - 2$

$= 3 + 16 - 2$

$= 19 - 2$

$= 17$

35. $100 \div 10 \cdot 10 \div 100 = 10 \cdot 10 \div 100$

$= 100 \div 100$

$= 1$

37. $15 - 2^2 + 2^3 = 15 - 4 + 8$

$= 11 + 8$

$= 19$

39. $3 + [7 + 3(2 + 5)] = 3 + [7 + 3(7)]$

$= 3 + [7 + 21]$

$= 3 + [28]$

$= 31$

41. $3(2[10 - 2(4 - 2)])$

$= 3(2[10 - 2(2)])$

$= 3(2[10 - 4])$

$= 3(2[6])$

$= 36$

43. $\dfrac{2}{5} + \dfrac{3}{5} \cdot 20 = \dfrac{2}{5} + \dfrac{60}{5}$

$= \dfrac{62}{5}$

$= 12\dfrac{2}{5}$

45. $(-2)(-4) + 5(-3) = 8 + (-15)$

$= -7$

47. $1 - 8 - 3 - 5 = 1 + (-8) + (-3) + (-5)$

$= 1 + (-16)$

$= -15$

49. $4 + 3[1 - 2(-5)]$

$= 4 + 3[1 + 10]$

$= 4 + 3[11]$

$= 4 + 33$

$= 37$

51. $2 + [4(-5 + 3) - 2]$

$= 2 + [4(-2) - 2]$

$= 2 + [-8 - 2]$

$= 2 + [-10]$

$= -8$

53. $-4^2 + 7(3[4 - 5 \cdot 2])$

$= -16 + 7(3[4 - 10])$

$= -16 + 7(3[-6])$

$= -16 + 7 (-18)$

$= -16 + (-126)$

$= -142$

55. $\left(-\dfrac{1}{2}\right)^2 \div \left(\dfrac{1}{2} - \dfrac{1}{3}\right)$

$= \dfrac{1}{4} \div \left(\dfrac{3}{6} - \dfrac{2}{6}\right)$

$= \dfrac{1}{4} \div \dfrac{1}{6}$

$= \dfrac{1}{\underset{2}{\cancel{4}}} \cdot \dfrac{\overset{3}{\cancel{6}}}{1}$

$= \dfrac{3}{2}$

57. $\left| \dfrac{-50}{6 - 11} \right| = \left| \dfrac{-50}{-5} \right| = |10| = 10$

59. $\dfrac{2(-12) + 4}{|-5 + 1|} = \dfrac{-24 + 4}{|-4|} = \dfrac{-20}{4} = -5$

61. $\dfrac{5(-2) + (-1)4}{-6 - 1} = \dfrac{-10 + (-4)}{-6 + (-1)} = \dfrac{-14}{-7} = 2$

63. $\dfrac{2 \cdot 3^2 - 3 \cdot 4^2}{2^2 + 1^2} = \dfrac{2 \cdot 9 - 3 \cdot 16}{4 + 1} = \dfrac{18 - 48}{5} = \dfrac{-30}{5} = -6$

65. $\dfrac{8[3 \cdot 5 - 4(3 \cdot 2 - 5)]}{4[5(1 + 6) - 4^2 + 9]}$

$= \dfrac{8[15 - 4(6 - 5)]}{4[5(7) - 16 + 9]} = \dfrac{8[15 - 4(1)]}{4[35 - 16 + 9]} = \dfrac{8[15 - 4]}{4[19 + 9]} = \dfrac{8[11]}{4[28]} = \dfrac{88}{112} = \dfrac{11}{14}$

67. $2x + 5 = 2(2) + 5 = 4 + 5 = 9$

69. $3x - 5y + 8 = 3(2) - 5(-3) + 8 = 6 - (-15) + 8 = 6 + 15 + 8 = 29$

71. $8x^2 - y^2 + 1 = 8(2)^2 - (-3)^2 + 1 = 8 \cdot 4 - 9 + 1 = 32 - 9 + 1 = 23 + 1 = 24$

73. $(-2y)(9x - 5z)^2 = -2(-3)(9 \cdot 2 - 5 \cdot 5)^2 = 6(18 - 25)^2 = 6(-7)^2 = 6 \cdot 49 = 294$

75. $\dfrac{-x^2 + 5x - 8}{|z - y^2|} = \dfrac{-2^2 + 5 \cdot 2 - 8}{|5 - (-3)^2|} = \dfrac{-4 + 10 - 8}{|5 - 9|} = \dfrac{6 - 8}{|-4|} = \dfrac{-2}{4} = -\dfrac{1}{2}$

77. $16 - 4(5) = 16 - 20 = \$-4$

79. $125 - 13 + 2 = 112 + 2 = 114$ lb

81. $3 \cdot 50 - 120 = 150 - 120 = \$30$

83. $2(-50) + 3(-35) + 10(200) = -100 + (-105) + 2000 = \$1795$

85. $\boxed{\text{Clear}}$ 7 $\boxed{x^2}$ 49

    $7^2 = 49$

87. $\boxed{\text{Clear}}$ 2 $\boxed{y^x}$ 20 $\boxed{=}$ 1,048,576

    $2^{20} = 1,048,576$

89. $\boxed{\text{Clear}}$ 4 $\boxed{+}$ 12 $\boxed{\div}$ 2 $\boxed{+}$ 2 $\boxed{=}$ 12

    $4 + 12 \div 2 + 2 = 12$

Problem Set 1.9, pp. 53-55

1. $-2 + 6 = 6 + (-2)$

3. $5(-7) = (-7)5$

5. $3 + x = x + 3$

7. $(1 + 2) + (-5) = 1 + (2 + (-5))$

9. $(-3 \cdot 4)5 = -3(4 \cdot 5)$

11. $\frac{1}{3}(3y) = (\frac{1}{3} \cdot 3)y$

13. $5 + 0 = 5$

15. $0 + x = x$

17. $1 \cdot y = y$

19. $2 + (-2) = 0$

21. $-8.2 + 8.2 = 0$

23. $\frac{3}{8} \cdot \frac{8}{3} = 1$

25. $3(5 + x) = 3 \cdot 5 + 3 \cdot x$

27. $(2 + 4)7 = 2 \cdot 7 + 4 \cdot 7$

29. $2(8y - 1) = 2(8y) - 2(1)$

31. $-1(x + y - 6) = (-1)x + (-1)y - (-1)6$

33. $2(5x) = (2 \cdot 5)x = 10x$

35. $\frac{3}{4}(\frac{4}{3}x) = (\frac{3}{4} \cdot \frac{4}{3})x = 1x = x$

37. $-8 + (x + 8) = -8 + (8 + x)$

$$= (-8 + 8) + x$$

$$= 0 + x$$

$$= x$$

39. $(7x)\frac{1}{7} = \frac{1}{7}(7x)$

$$= (\frac{1}{7} \cdot 7)x$$

$$= 1x$$

$$= x$$

41. $4(x + 1) = 4 \cdot x + 4 \cdot 1$

$$= 4x + 4$$

43. $6(2y + 4) = 6(2y) + 6(4)$

$$= 12y + 24$$

45. $8(3x - 4) = 8(3x) - 8(4)$

$$= 24x - 32$$

47. $16(\frac{3t}{2} - \frac{5}{8}) = 16(\frac{3t}{2}) - 16(\frac{5}{8})$

$$= 24t - 10$$

49. $2(2x + 3y - 4) = 2(2x) + 2(3y) - 2(4) = 4x + 6y - 8$

51. $(x + 7) + (-3) = x + (7 + (-3)) = x + 4$

53. $x + (-x + 5) = (x + (-x)) + 5 = 0 + 5 = 5$

55. $(x + 1) + (-x) = (1 + x) + (-x) = 1 + (x + (-x)) = 1 + 0 = 1$

57. $x + 5 - 1 = x + 5 + (-1) = x + 4$

59. $x - 8 + 8 = x + (-8) + 8 = x + 0 = x$

61. $y - 2 - 3 = y + (-2) + (-3) = y + (-5) = y - 5$

63. $4 + y - 5 = 4 + y + (-5) = y + 4 + (-5) = y + (-1) = y - 1$

65. $-2(3x) = (-2 \cdot 3)x = -6x$

67. $5(-4x) = [5(-4)]x = -20x$

69. $(-\frac{3}{5}y)10 = 10(-\frac{3}{5}y) = [10(-\frac{3}{5})]y = -6y$

71. $-\frac{1}{2}(-2r) = [(-\frac{1}{2})(-2)]r = 1r = r$

73. $-3(2x + 5) = -3(2x) + (-3)5 = -6x + (-15) = -6x - 15$

75. $5(-4x + 1) = 5(-4x) + 5(1) = -20x + 5$

77. $-(y - 7) = -1(y - 7) = (-1)y - (-1)7 = -y - (-7) = -y + 7$

79. $-(8y + 12) = -1(8y + 12) = (-1)8y + (-1)12 = -8y + (-12) = -8y - 12$

81. $-(-a - 3b + 5c) = -1(-a - 3b + 5c) = (-1)(-a) - (-1)3b + (-1)5c$

$$= a - (-3b) + (-5c)$$

$$= a + 3b - 5c$$

83. $-4(\frac{x}{2} - \frac{y}{4} + 3) = (-4)\frac{x}{2} - (-4)\frac{y}{4} + (-4)3 = -2x - (-y) + (-12)$

$$= -2x + y - 12$$

85. $\frac{-15x}{3} = \frac{-15}{3} \cdot \frac{x}{1} = -5x$

87. $\frac{8x}{-4} = \frac{8}{-4} \cdot \frac{x}{1} = -2x$

89. $\frac{-3y}{-3} = \frac{-3}{-3} \cdot \frac{y}{1} = 1y = y$

91. $5 - 2 = 3$ but $2 - 5 = -3$.

93. $6 - (4 - 1) = 6 - 3 = 3$ but $(6 - 4) - 1 = 2 - 1 = 1$.

NOTES

CHAPTER 2

LINEAR EQUATIONS AND INEQUALITIES

Problem Set 2.1, pp. 65-66

1. x, 3

3. -3x, 4y, -10

5. x^2, 5xy, $-y^2$

7. 4

9. 6

11. -15

13. 1

15. $\frac{2}{3}$

17. 2.9

19. Like

21. Unlike

23. Unlike

25. Like

27. Like

29. Like

31. $5x + 4x = (5 + 4)x = 9x$

33. $17z + 24z - 1 = 41z - 1$

35. $8m + 11m + 13m = 32m$

37. $2y^2 + 10y^2 + y^2$

$= (2 + 10 + 1)y^2$

$= 13y^2$

39. $7p - p = (7 - 1)p$

$= 6p$

41. $9p^2 - p^2 + p = 8p^2 + p$

43. $-11rs + 9rs = -2rs$

45. $5m - 10m = -5m$

47. $4x - 3 - 3x = x - 3$

49. $-4y^2 - y^2 = -5y^2$

51. $-6r + 6r - 1 = 0 - 1 = -1$

53. $-3z^3 + 4z^3 = 1z^3 = z^3$

55. $2t^3 + 5t^2 + 5t^3 = 7t^3 + 5t^2$

57. $-2y - 4 + 4 + 2y = 0$

59. $8m + 15 + 3m = 11m + 15$

61. $-3r + 9 + 8r - 9 = 5r$

63. $6p + 4 - 2p - 19 = 4p - 15$

65. $-t - 5 - 4t - 5 = -5t - 10$

67. $15y - 13x - 2y + 2x = -11x + 13y$

69. $-9r^2 + 4r - r^2 - 2r + 6 = -10r^2 + 2r + 6$

71. $8 + 2(x + 3) = 8 + 2x + 6 = 2x + 14$

73. $6y - (y + 5) = 6y - y - 5 = 5y - 5$

75. $15y - 4(5y - 2) = 15y - 20y + 8 = -5y + 8$

77. $5(2x - 1) - 3(-x + 7) = 10x - 5 + 3x - 21 = 13x - 26$

79. $7(x - 4) - (x - 6) = 7x - 28 - x + 6 = 6x - 22$

81. $3(a - 2b) - 7(3a - b) = 3a - 6b - 21a + 7b = -18a + b$

83. $-2(3x - 4y) - (x + 8y) = -6x + 8y - x - 8y = -7x$

85. $-(11x + 9y) - (x - 14y) = -11x - 9y - x + 14y = -12x + 5y$

87. $5 + 5(p + 1) - 6(2p) = 5 + 5p + 5 - 12p = -7p + 10$

89. $4 + 2(m - 13) + 3(-3m) = 4 + 2m - 26 + (-9m) = -7m - 22$

91. $2 + 4(a - 1) + 3(-5a) - (-2a)4 = 2 + 4a - 4 - 15a + 8a = -3a - 2$

93. $3 + 6[k + 2(k + 1)] = 3 + 6[k + 2k + 2]$

$$= 3 + 6[3k + 2]$$

$$= 3 + 18k + 12$$

$$= 18k + 15$$

95. $12 - 2[m - 3(2m - 5)] = 12 - 2[m - 6m + 15]$

$$= 12 - 2[-5m + 15]$$

$$= 12 + 10m - 30$$

$$= 10m - 18$$

97. $1 + 4[6x - (5x + 4) - x] = 1 + 4[6x - 5x - 4 - x]$

$$= 1 + 4[-4]$$

$$= 1 - 16$$

$$= -15$$

Problem Set 2.2, pp. 70-71

1. $2y - 5 = 1$

$2(3) - 5 = 1$

$6 - 5 = 1$

$1 = 1$ True

3 is a solution.

3. $4m + 9 = 11$

$4(-5) + 9 = 11$

$-20 + 9 = 11$

$-11 = 11$ False

-5 is not a solution.

5. $3p - 5p - 2 = 0$

$3(-1) - 5(-1) - 2 = 0$

$-3 + 5 - 2 = 0$

$0 = 0$ True

-1 is a solution.

7. $6(t - 5) = 3t - 24$

$6(2 - 5) = 3(2) - 24$

$6(-3) = 6 - 24$

$-18 = -18$ True

2 is a solution.

9. $x - 3 = 4$

$x - 3 + 3 = 4 + 3$

$x = 7$

11. $y + 7 = 3$

$y + 7 - 7 = 3 - 7$

$y = -4$

13. $k - 5 = -11$

$k - 5 + 5 = -11 + 5$

$k = -6$

15. $61 + p = 104$

$61 + p - 61 = 104 - 61$

$p = 43$

17. $7.8 + m = -5.3$

$7.8 + m - 7.8 = -5.3 - 7.8$

$m = -13.1$

19. $-19 = t - 19$

$-19 + 19 = t - 19 + 19$

$0 = t$

21. $r + 23 = 0$

$r + 23 - 23 = 0 - 23$

$r = -23$

23. $\frac{1}{2} = \frac{1}{2} + z$

$\frac{1}{2} - \frac{1}{2} = \frac{1}{2} + z - \frac{1}{2}$

$0 = z$

25. $k - \dfrac{1}{2} = -\dfrac{5}{2}$

$k - \dfrac{1}{2} + \dfrac{1}{2} = -\dfrac{5}{2} + \dfrac{1}{2}$

$k = -\dfrac{4}{2}$

$k = -2$

27. $5x = 4x + 6$

$5x - 4x = 4x + 6 - 4x$

$x = 6$

29. $8y - 2 = 9y$

$8y - 2 - 8y = 9y - 8y$

$-2 = y$

31. $15 - 7k = -6k$

$15 - 7k + 7k = -6k + 7k$

$15 = k$

33. $-2m = -10 - 3m$

$-2m + 3m = -10 - 3m + 3m$

$m = -10$

35. $2t + 3 = t - 8$

$2t + 3 - t = t - 8 - t$

$t + 3 = -8$

$t + 3 - 3 = -8 - 3$

$t = -11$

37. $5p + 7 = 6p - 4$

$5p + 7 - 5p = 6p - 4 - 5p$

$7 = p - 4$

$7 + 4 = p - 4 + 4$

$11 = p$

39. $-14 - 10r = -9r - 1$

$-14 - 10r + 10r = -9r - 1 + 10r$

$-14 = r - 1$

$-14 + 1 = r - 1 + 1$

$-13 = r$

41. $5.4 - 2.7z = 9.9 - 3.7z$

$5.4 - 2.7z + 3.7z = 9.9 - 3.7z + 3.7z$

$5.4 + z = 9.9$

$5.4 + z - 5.4 = 9.9 - 5.4$

$z = 4.5$

43. $$\frac{4}{3}x + \frac{2}{5} = \frac{1}{3}x + \frac{9}{5}$$

$$\frac{4}{3}x + \frac{2}{5} - \frac{1}{3}x = \frac{1}{3}x + \frac{9}{5} - \frac{1}{3}x$$

$$x + \frac{2}{5} = \frac{9}{5}$$

$$x + \frac{2}{5} - \frac{2}{5} = \frac{9}{5} - \frac{2}{5}$$

$$x = \frac{7}{5}$$

45. $6x + 3 - 2x - 3x = 4 + 8$

$x + 3 = 12$

$x + 3 - 3 = 12 - 3$

$x = 9$

47. $5y - 9 + y = 3y - 7 + 2y$

$6y - 9 = 5y - 7$

$6y - 9 - 5y = 5y - 7 - 5y$

$y - 9 = -7$

$y - 9 + 9 = -7 + 9$

$y = 2$

49. $-4p + 6p + 2 - 8 = 2p - 13 + p$

$2p - 6 = 3p - 13$

$2p - 6 - 2p = 3p - 13 - 2p$

$-6 = p - 13$

$-6 + 13 = p - 13 + 13$

$7 = p$

51. $-12 - 3 - 9z - 1 + 4z = -3z - 4 - 3z$

$-16 - 5z = -6z - 4$

$-16 - 5z + 6z = -6z - 4 + 6z$

$-16 + z = -4$

$-16 + z + 16 = -4 + 16$

$z = 12$

53. $3(2m - 4) - 5m = 3 + 9$

$6m - 12 - 5m = 12$

$m - 12 = 12$

$m - 12 + 12 = 12 + 12$

$m = 24$

55. $4(t + 2) = 3t - 6$

$4t + 8 = 3t - 6$

$4t + 8 - 3t = 3t - 6 - 3t$

$t + 8 = -6$

$t + 8 - 8 = -6 - 8$

$t = -14$

57. $-8(k + 3) = 14 - 7k$

$-8k - 24 = 14 - 7k$

$-8k - 24 + 8k = 14 - 7k + 8k$

$-24 = 14 + k$

$-24 - 14 = 14 + k - 14$

$-38 = k$

59. $2y - (y - 5) = -8 - 4$

$2y - y + 5 = -12$

$y + 5 = -12$

$y + 5 - 5 = -12 - 5$

$y = -17$

61. $5(3 - 2x) + 3(7 + 2x) = 9 - 5x$

$15 - 10x + 21 + 6x = 9 - 5x$

$36 - 4x = 9 - 5x$

$36 - 4x + 5x = 9 - 5x + 5x$

$36 + x = 9$

$36 + x - 36 = 9 - 36$

$x = -27$

63. $9 - 4(3r - 6) = 10r - (23r + 1)$

$9 - 12r + 24 = 10r - 23r - 1$

$33 - 12r = -13r - 1$

$33 - 12r + 13r = -13r - 1 + 13r$

$33 + r = -1$

$33 + r - 33 = -1 - 33$

$r = -34$

65. $12(\frac{x}{6} + \frac{2}{3}) = 7x - 6x + 8$

$2x + 8 = x + 8$

$2x + 8 - x = x + 8 - x$

$x + 8 = 8$

$x + 8 - 8 = 8 - 8$

$x = 0$

67. $\frac{2}{3} + \frac{3z}{8} = \frac{2}{9} - \frac{5z}{8}$

$\frac{2}{3} + \frac{3z}{8} + \frac{5z}{8} = \frac{2}{9} - \frac{5z}{8} + \frac{5z}{8}$

$\frac{2}{3} + \frac{8z}{8} = \frac{2}{9}$

$\frac{2}{3} + z - \frac{2}{3} = \frac{2}{9} - \frac{2}{3}$

$z = -\frac{4}{9}$

Problem Set 2.3, pp. 74-75

1. $3x = 6$

$$\frac{3x}{3} = \frac{6}{3}$$

$$x = 2$$

3. $12y = 3$

$$\frac{12y}{12} = \frac{3}{12}$$

$$y = \frac{1}{4}$$

5. $-4y = 20$

$$\frac{-4y}{-4} = \frac{20}{-4}$$

$$y = -5$$

7. $-m = 4$

$$(-1)(-m) = (-1)4$$

$$m = -4$$

9. $-3z = -2$

$$\frac{-3z}{-3} = \frac{-2}{-3}$$

$$z = \frac{2}{3}$$

11. $\frac{2}{3}p = 6$

$$\frac{3}{2}\cdot\frac{2}{3}p = \frac{3}{2}\cdot 6$$

$$p = 9$$

13. $\frac{5}{2}p = 3$

$$\frac{2}{5}\cdot\frac{5}{2}p = \frac{2}{5}\cdot 3$$

$$p = \frac{6}{5}$$

15. $-24 = -\frac{3}{4}r$

$$-\frac{4}{3}(-24) = -\frac{4}{3}(-\frac{3}{4}r)$$

$$32 = r$$

17. $-2x = 0$

$$\frac{-2x}{-2} = \frac{0}{-2}$$

$$x = 0$$

19. $\frac{m}{8} = -2$

$$8(\frac{m}{8}) = 8(-2)$$

$$m = -16$$

21. $7y = \frac{7}{9}$

$$\frac{1}{7}\cdot 7y = \frac{1}{7}\cdot\frac{7}{9}$$

$$y = \frac{1}{9}$$

23. $-\frac{5}{3} = 3k$

$$\frac{1}{3}(-\frac{5}{3}) = \frac{1}{3}(3k)$$

$$-\frac{5}{9} = k$$

25. $\frac{2}{7}x = 0$

$\frac{7}{2} \cdot \frac{2}{7}x = \frac{7}{2} \cdot 0$

$x = 0$

27. $-z = \frac{3}{8}$

$(-1)(-z) = (-1)\frac{3}{8}$

$z = -\frac{3}{8}$

29. $\frac{5}{4}y = -\frac{5}{12}$

$\frac{4}{5}(\frac{5}{4}y) = \frac{4}{5}(-\frac{5}{12})$

$y = -\frac{1}{3}$

31. $-\frac{r}{2} = \frac{4}{5}$

$-2(-\frac{r}{2}) = -2(\frac{4}{5})$

$r = -\frac{8}{5}$

33. $5.6t = -44.8$

$\frac{5.6t}{5.6} = \frac{-44.8}{5.6}$

$t = -8$

35. $-1.08z = -7.02$

$\frac{-1.08z}{-1.08} = \frac{-7.02}{-1.08}$

$z = 6.5$

37. $5x - 2x + 6x = 4 + 12 + 2$

$9x = 18$

$\frac{9x}{9} = \frac{18}{9}$

$x = 2$

39. $-3 + 4r - 7r + 3 = 2 - 5$

$-3r = -3$

$\frac{-3r}{-3} = \frac{-3}{-3}$

$r = 1$

41. $3m + 8 - 3m = -10m + m + 5m$

$8 = -4m$

$\frac{8}{-4} = \frac{-4m}{-4}$

$-2 = m$

43. $-8k + 12k - 2k = -2 - 3 - 9$

$2k = -14$

$\frac{2k}{2} = \frac{-14}{2}$

$k = -7$

45. $-9p - 9p - 7 + 7 = 12 + 6$

$$-18p = 18$$

$$\frac{-18p}{-18} = \frac{18}{-18}$$

$$p = -1$$

47. $3z - 3z + z - 2z = -8 + 3$

$$-z = -5$$

$$(-1)(-z) = (-1)(-5)$$

$$z = 5$$

49. $15r - 8r - 8r = r - 6r + 5r$

$$-r = 0$$

$$(-1)(-r) = (-1)0$$

$$r = 0$$

51. $2(x - 3) + 6 = 12 + 8$

$$2x - 6 + 6 = 20$$

$$2x = 20$$

$$\frac{2x}{2} = \frac{20}{2}$$

$$x = 10$$

53. $4(3y + 1) - (8y + 4) = -1 - 1$

$$12y + 4 - 8y - 4 = -2$$

$$4y = -2$$

$$\frac{4y}{4} = \frac{-2}{4}$$

$$y = -\frac{1}{2}$$

55. $-3(6 - 2k) + 2(k + 9) = -15 - 9$

$$-18 + 6k + 2k + 18 = -24$$

$$8k = -24$$

$$\frac{8k}{8} = \frac{-24}{8}$$

$$k = -3$$

57. $5(m + 4) - 5m = -2m - 3(m - 1) - 3$

$$5m + 20 - 5m = -2m - 3m + 3 - 3$$

$$20 = -5m$$

$$\frac{20}{-5} = \frac{-5m}{-5}$$

$$-4 = m$$

59. $6.28k - 2.08k = 18.3 + 2.7$

$4.2k = 21$

$\dfrac{4.2k}{4.2} = \dfrac{21}{4.2}$

$k = 5$

61. $4\left(\dfrac{x}{2} - \dfrac{3}{4}\right) + 3 = 32 - 7 - 1$

$4\cdot\dfrac{x}{2} - 4\cdot\dfrac{3}{4} + 3 = 24$

$2x - 3 + 3 = 24$

$2x = 24$

$\dfrac{2x}{2} = \dfrac{24}{2}$

$x = 12$

63. $\dfrac{2}{3}p + \dfrac{2}{3}p = \dfrac{2}{3} + \dfrac{1}{2}$

$\dfrac{4}{3}p = \dfrac{7}{6}$ Since $\dfrac{2}{3} + \dfrac{1}{2} = \dfrac{4}{6} + \dfrac{3}{6} = \dfrac{7}{6}$

$\dfrac{3}{4}\cdot\dfrac{4}{3}p = \dfrac{3}{4}\cdot\dfrac{7}{6}$

$p = \dfrac{7}{8}$ Since $\dfrac{3}{4}\cdot\dfrac{7}{6} = \dfrac{21}{24} = \dfrac{7}{8}$

Problem Set 2.4, pp. 79-80

1. $5x - 3 = 12$

$5x - 3 + 3 = 12 + 3$

$5x = 15$

$\dfrac{5x}{5} = \dfrac{15}{5}$

$x = 3$

3. $2p + 5 = -13$

$2p + 5 - 5 = -13 - 5$

$2p = -18$

$\dfrac{2p}{2} = \dfrac{-18}{2}$

$p = -9$

5. $2t + 11 = 11$

$2t + 11 - 11 = 11 - 11$

$2t = 0$

$\dfrac{2t}{2} = \dfrac{0}{2}$

$t = 0$

7. $4.2r - 1.7 = -12.2$

$4.2r - 1.7 + 1.7 = -12.2 + 1.7$

$4.2r = -10.5$

$\dfrac{4.2r}{4.2} = \dfrac{-10.5}{4.2}$

$r = -2.5$

9. $-\frac{3}{2}m + 5 = -19$

$-\frac{3}{2}m + 5 - 5 = -19 - 5$

$-\frac{3}{2}m = -24$

$-\frac{2}{3}(-\frac{3}{2}m) = -\frac{2}{3}(-24)$

$m = 16$

11. $-7k + 8 = 34$

$-7k + 8 - 8 = 34 - 8$

$-7k = 26$

$\frac{-7k}{-7} = \frac{26}{-7}$

$k = -\frac{26}{7}$

13. $6 - x = 12$

$6 - x - 6 = 12 - 6$

$-x = 6$

$(-1)(-x) = (-1)6$

$x = -6$

15. $15 - 2y = 25$

$15 - 2y - 15 = 25 - 15$

$-2y = 10$

$\frac{-2y}{-2} = \frac{10}{-2}$

$y = -5$

17. $-19 - 4p = -3$

$-19 - 4p + 19 = -3 + 19$

$-4p = 16$

$\frac{-4p}{-4} = \frac{16}{-4}$

$p = -4$

19. $2x + 5 = 9 + 4$

$2x + 5 = 13$

$2x + 5 - 5 = 13 - 5$

$2x = 8$

$\frac{2x}{2} = \frac{8}{2}$

$x = 4$

21. $5y + 3y = -24$

$8y = -24$

$\frac{8y}{8} = \frac{-24}{8}$

$y = -3$

23. $8p - 9p = -10 - 1$

$-p = -11$

$(-1)(-p) = (-1)(-11)$

$p = 11$

25. $\frac{11}{3}r - \frac{7}{3}r + 3 = 15$

$\frac{4}{3}r + 3 = 15$

$\frac{4}{3}r + 3 - 3 = 15 - 3$

$\frac{4}{3}r = 12$

$\frac{3}{4} \cdot \frac{4}{3}r = \frac{3}{4} \cdot 12$

$r = 9$

27. $x - 6x + 4 = 20 - 21$

$-5x + 4 = -1$

$-5x + 4 - 4 = -1 - 4$

$-5x = -5$

$\frac{-5x}{-5} = \frac{-5}{-5}$

$x = 1$

29. $9z + 1 = -6z - 29$

$9z + 1 + 6z = -6z - 29 + 6z$

$15z + 1 = -29$

$15z + 1 - 1 = -29 - 1$

$15z = -30$

$\frac{15z}{15} = \frac{-30}{15}$

$z = -2$

31. $-p + 14 = p + 9$

$-p + 14 + p = p + 9 + p$

$14 = 2p + 9$

$14 - 9 = 2p + 9 - 9$

$5 = 2p$

$\frac{5}{2} = \frac{2p}{2}$

$\frac{5}{2} = p$

33. $r + 7 = 2r - 6$

$r + 7 - r = 2r - 6 - r$

$7 = r - 6$

$7 + 6 = r - 6 + 6$

$13 = r$

35. $7t - 4 = t - 4$

$7t - 4 - t = t - 4 - t$

$6t - 4 = -4$

$6t - 4 + 4 = -4 + 4$

$6t = 0$

$\frac{6t}{6} = \frac{0}{6}$

$t = 0$

37. $5(x - 1) = 3x - 7$

$5x - 5 = 3x - 7$

$5x - 5 - 3x = 3x - 7 - 3x$

$2x - 5 = -7$

$2x - 5 + 5 = -7 + 5$

$2x = -2$

$\dfrac{2x}{2} = \dfrac{-2}{2}$

$x = -1$

39. $3y - 2(y + 1) = 5(y + 2)$

$3y - 2y - 2 = 5y + 10$

$y - 2 = 5y + 10$

$y - 2 - y = 5y + 10 - y$

$-2 = 4y + 10$

$-2 - 10 = 4y + 10 - 10$

$-12 = 4y$

$\dfrac{-12}{4} = \dfrac{4y}{4}$

$-3 = y$

41. $3(k + 5) + 4(k + 5) = 21$

$3k + 15 + 4k + 20 = 21$

$7k + 35 = 21$

$7k + 35 - 35 = 21 - 35$

$7k = -14$

$\dfrac{7k}{7} = \dfrac{-14}{7}$

$k = -2$

43. $m - (4m - 5) = -19$

$m - 4m + 5 = -19$

$-3m + 5 = -19$

$-3m + 5 - 5 = -19 - 5$

$-3m = -24$

$\dfrac{-3m}{-3} = \dfrac{-24}{-3}$

$m = 8$

45. $8 - (w - 4) = 3w + 9$

$8 - w + 4 = 3w + 9$

$12 - w = 3w + 9$

$12 - w + w = 3w + 9 + w$

$12 = 4w + 9$

$12 - 9 = 4w + 9 - 9$

$3 = 4w$

$\dfrac{3}{4} = \dfrac{4w}{4}$

$\dfrac{3}{4} = w$

47. $15 - 3(a + 6) = 9 - (4a + 5)$

$15 - 3a - 18 = 9 - 4a - 5$

$-3a - 3 = 4 - 4a$

$-3a - 3 + 4a = 4 - 4a + 4a$

$-3 + a = 4$

$-3 + a + 3 = 4 + 3$

$a = 7$

49. $t - 2(3t - 2) = -t - 16$

$t - 6t + 4 = -t - 16$

$-5t + 4 = -t - 16$

$-5t + 4 + 5t = -t - 16 + 5t$

$4 = 4t - 16$

$4 + 16 = 4t - 16 + 16$

$20 = 4t$

$\dfrac{20}{4} = \dfrac{4t}{4}$

$5 = t$

51. $7 - (3p + 2) = 2p + 5(p + 7)$

$7 - 3p - 2 = 2p + 5p + 35$

$5 - 3p = 7p + 35$

$5 - 3p + 3p = 7p + 35 + 3p$

$5 = 10p + 35$

$5 - 35 = 10p + 35 - 35$

$-30 = 10p$

$\dfrac{-30}{10} = \dfrac{10p}{10}$

$-3 = p$

53. $6.1k - 4.3(2k - 5) = 2(1.5 + k) - 10.3$

$6.1k - 8.6k + 21.5 = 3 + 2k - 10.3$

$-2.5k + 21.5 = 2k - 7.3$

$-2.5k + 21.5 + 2.5k = 2k - 7.3 + 2.5k$

$21.5 = 4.5k - 7.3$

$21.5 + 7.3 = 4.5k - 7.3 + 7.3$

$28.8 = 4.5k$

$\dfrac{28.8}{4.5} = \dfrac{4.5k}{4.5}$

$6.4 = k$

55. $-\dfrac{1}{3}(6x + 3) - 1 = \dfrac{1}{2}(4x - 10)$

$-2x - 1 - 1 = 2x - 5$

$-2x - 2 = 2x - 5$

$-2x - 2 + 2x = 2x - 5 + 2x$

$-2 = 4x - 5$

$-2 + 5 = 4x - 5 + 5$

$3 = 4x$

$\dfrac{3}{4} = \dfrac{4x}{4}$

$\dfrac{3}{4} = x$

57. $-2[8y - 5(2y - 3)] = 3y - 8$

$-2[8y - 10y + 15] = 3y - 8$

$-2[-2y + 15] = 3y - 8$

$4y - 30 = 3y - 8$

$4y - 30 - 3y = 3y - 8 - 3y$

$y - 30 = -8$

$y - 30 + 30 = -8 + 30$

$y = 22$

59. $5 + 3[1 + 2(2m - 3)] = 6(m + 5)$

$5 + 3[1 + 4m - 6] = 6m + 30$

$5 + 3[4m - 5] = 6m + 30$

$5 + 12m - 15 = 6m + 30$

$-10 + 12m = 6m + 30$

$-10 + 12m - 6m = 6m + 30 - 6m$

$-10 + 6m = 30$

$-10 + 6m + 10 = 30 + 10$

$6m = 40$

$\dfrac{6m}{6} = \dfrac{40}{6}$

$m = \dfrac{20}{3}$

Problem Set 2.5, pp. 86-89

1. $P = 2\ell + 2w$

$P = 2(6) + 2(5)$

$P = 12 + 10$

$P = 22$ yd

3. $P = 2\ell + 2w$

$82 = 2\ell + 2(16)$

$82 = 2\ell + 32$

$50 = 2\ell$

25 ft $= \ell$

5. $A = \ell w$

$A = 4 \cdot 3$

$A = 12$ sq m

7. $A = \ell w$

$36 = 8w$

4.5 cm $= w$

9. $A = \pi r^2$

$A = 3.14(10)^2$

$A = 3.14(100)$

$A = 314$ sq in.

11. $A = \pi r^2$

$A = 3.14(\frac{1}{2})^2$

$A = 3.14(\frac{1}{4})$

$A = 0.785$ sq mi

13. $C = \pi d$

    $C = 3.14(5)$

    $C = 15.7$ ft

15. $C = \pi d$

    $15.7 = 3.14d$

    5 km $= d$

17. $V = \ell wh$

    $V = (22)(9)(5)$

    $V = 990$ cu in.

19. $V = \ell wh$

    $600 = (25)w(2)$

    $600 = 50w$

    12 ft $= w$

21. $I = Prt$

    $I = (375)(0.12)(1)$

    $I = \$45$

23. $I = Prt$

    $306 = P(0.09)(4)$

    $306 = 0.36P$

    $\$850 = P$

25. $d = rt$

    $d = (25)(3)$

    $d = 75$ mi

27. $d = rt$

    $d = (60)(\frac{1}{3})$ Since 20 min $= \frac{1}{3}$ hr

    $d = 20$ mi

29. $A = \frac{1}{2}bh$

    $A = (\frac{1}{2})(13)(8)$

    $A = 52$ sq cm

31. $A = \frac{1}{2}bh$

    $50 = (\frac{1}{2})b(5)$

    $50 = \frac{5}{2}b$

    $\frac{2}{5} \cdot 50 = \frac{2}{5} \cdot \frac{5}{2}b$

    20 yd $= b$

33. $d = rt$

$$\frac{d}{r} = \frac{rt}{r}$$

$$\frac{d}{r} = t \text{ or } t = \frac{d}{r}$$

(a) $t = \frac{420}{30}$ (b) $t = \frac{420}{35}$ (c) $t = \frac{420}{40}$

$t = 14$ hr $t = 12$ hr $t = 10.5$ hr

35. $A = \ell w$

$$\frac{A}{w} = \frac{\ell w}{w}$$

$$\frac{A}{w} = \ell$$

37. $I = Prt$

$$\frac{I}{Pt} = \frac{Prt}{Pt}$$

$$\frac{I}{Pt} = r$$

39. $P = 2\ell + 2w$

$P - 2\ell = 2\ell + 2w - 2\ell$

$P - 2\ell = 2w$

$$\frac{P - 2\ell}{2} = \frac{2w}{2}$$

$$\frac{P - 2\ell}{2} = w$$

41. $I = P + Prt$

$I - P = P + Prt - P$

$I - P = Prt$

$$\frac{I - P}{Pr} = \frac{Prt}{Pr}$$

$$\frac{I - P}{Pr} = t$$

43. $A = \frac{1}{2}bh$

$2 \cdot A = 2 \cdot \frac{1}{2}bh$

$2A = bh$

$$\frac{2A}{b} = \frac{bh}{b}$$

$$\frac{2A}{b} = h$$

45. $s = \frac{1}{2}(a + b + c)$

$2 \cdot s = 2 \cdot \frac{1}{2}(a + b + c)$

$2s = a + b + c$

$2s - b - c = a + b + c - b - c$

$2s - b - c = a$

47. $A = \frac{h}{2}(B + b)$

   $2 \cdot A = 2 \cdot \frac{h}{2}(B + b)$

   $2A = h(B + b)$

   $2A = hB + hb$

   $2A - hB = hB + hb - hB$

   $2A - hB = hb$

   $\frac{2A - hB}{h} = \frac{hb}{h}$

   $\frac{2A - hB}{h} = b$

49. $c^2 = a^2 + b^2$

   $c^2 - b^2 = a^2 + b^2 - b^2$

   $c^2 - b^2 = a^2$

51. $V = \frac{1}{3}\pi r^2 h$

   $3 \cdot V = 3 \cdot \frac{1}{3}\pi r^2 h$

   $3V = \pi r^2 h$

   $\frac{3V}{\pi h} = \frac{\pi r^2 h}{\pi h}$

   $\frac{3V}{\pi h} = r^2$

53. $C = \frac{5}{9}(F - 32)$

   $\frac{9}{5} \cdot C = \frac{9}{5} \cdot \frac{5}{9}(F - 32)$

   $\frac{9}{5}C = F - 32$

   $\frac{9}{5}C + 32 = F - 32 + 32$

   $\frac{9}{5}C + 32 = F$

   or $F = \frac{9}{5}C + 32$

(a) $F = \frac{9}{5}(75) + 32 = 135 + 32 = 167°$

(b) $F = \frac{9}{5}(0) + 32 = 0 + 32 = 32°$

(c) $F = \frac{9}{5}(-20) + 32 = -36 + 32 = -4°$

55. $r = \frac{d}{t}$

   $r = \frac{4700}{1/2}$ $(30 \min = \frac{1}{2} hr)$

   $r = 9400$ mph

57. $d = rt$

   $d = (1100)(4)$

   $d = 4400$ ft

59. Since the radius, r, is half the diameter, d = 8 in., the area of one small pizza is

$$A = \pi r^2 = 3.14(4^2) = 3.14(16) = 50.24 \text{ sq in.}$$

The area of two small pizzas is

$$A = 2(50.24) = 100.48 \text{ sq in.}$$

The area of one large pizza is

$$A = \pi r^2 = 3.14(6^2) = 3.14(36) = 113.04 \text{ sq in.}$$

Therefore, the large pizza is the better buy.

61. $V = \frac{1}{3} \pi r^2 h$

$$V = \frac{1}{3}(3.14)(4^2)(9) = (3.14)(16)(3) = 150.72 \text{ cu ft}$$

63. Together, the two curved ends form a circle whose circumference is

$$C = \pi d = 3.14(55) = 172.7 \text{ yd.}$$

Therefore, the total length of the track is

$$\ell = 172.7 + 130 + 130 = 432.7 \text{ yd.}$$

65. The area of the top rectangle is

$$A = \ell w = 10 \cdot 9 = 90 \text{ sq ft.}$$

The area of the bottom rectangle is

$$A = \ell w = 24 \cdot 15 = 360 \text{ sq ft.}$$

The total area is

$$A = 90 + 360 = 450 \text{ sq ft, or } A = \frac{450}{9} = 50 \text{ sq yd.}$$

Therefore, the cost is

$$C = 50 \cdot 15 = \$750.$$

67.   The area of the rectangle is

$$A = \ell w = 14 \cdot 6 = 84 \text{ sq cm.}$$

The area of the triangle is

$$A = \frac{1}{2}bh = \frac{1}{2}(6)(8) = 24 \text{ sq cm.}$$

The area of the circle is

$$A = \pi r^2 = 3.14(3^2) = 3.14(9) = 28.26 \text{ sq cm.}$$

Therefore, the area of the shaded region is

$$A = 84 - (24 + 28.26) = 84 - 52.26 = 31.74 \text{ sq cm.}$$

69.   The grazing area is three-fourths the area of a circle of radius

20 ft.

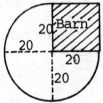

$$A = \frac{3}{4}\pi r^2$$

$$A = \frac{3}{4}(3.14)(20^2)$$

$$A = \frac{3}{4}(3.14)(400) = 3.14(300) = 942 \text{ sq ft}$$

71. $C = \pi d = \pi(2r) = 2\pi r = 2\pi(17.8)$

| Clear | 2 | x | π | x | 17.8 | = | 111.84070 |

$C = 2\pi(17.8) \approx 111.841 \text{ cm}$

Problem Set 2.6, pp. 94-96

1. $2x + 5$

3. $2(x + 5)$

5. $\frac{x}{3} - 1$

7. $6x^2$

9. $-8x$

11. $\frac{x}{2} - \frac{1}{2}$

13. $60t$ sec

15. $175(z + 6)$ cal

17. $5x + 25y$ cents

19. $0.11m$ dollars

21. $0.05(40 - p)$ ¢

23. $\dfrac{60}{x}$ ft

25. $x, x + 1, x + 2$

27. $x, x + 2, x + 4, x + 6$

29. $\dfrac{2}{3}x, x$

31. $x, x + 6$

33. $x, x - 5, 2x, 2x - 5$

35. $x, 500 - x$

37. $2(x + 8) + x = 2x + 16 + x = 3x + 16$

39. $x + 1 - 4(x - 7) = x + 1 - 4x + 28 = -3x + 29$

41. $3(x + x + 2) - 6 = 3(2x + 2) - 6 = 6x + 6 - 6 = 6x$

43. $x - 0.08x = 1.00x - 0.08x = 0.92x$

45. $x + (x + 10) + 2(x + 10) = x + x + 10 + 2x + 20 = 4x + 30$ hr

47. $x =$ the number

$$2x + 7 = 29$$
$$2x = 22$$
$$x = 11$$

49. $x =$ the number

$$\frac{3}{5}x = 45$$
$$\frac{5}{3} \cdot \frac{3}{5}x = \frac{5}{3} \cdot 45$$
$$x = 75$$

51. $x =$ the number

$$0.38x = 20.9$$
$$\frac{0.38x}{0.38} = \frac{20.9}{0.38}$$
$$x = 55$$

53. $x =$ the number

$$3x = 10 + 2x$$
$$3x - 2x = 10 + 2x - 2x$$
$$x = 10$$

55. x = the number

$4x - 7 = x - 1$

$3x - 7 = -1$

$3x = 6$

$x = 2$

57. x = the number

$9(x + 4) = x - 4$

$9x + 36 = x - 4$

$8x + 36 = -4$

$8x = -40$

$x = -5$

59. x = the cost

$x + 0.30x = 49.40$

$1.3x = 49.40$

$\dfrac{1.3x}{1.3} = \dfrac{49.40}{1.3}$

$x = \$38$

61. x = original population

$x - 0.07x = 5115$

$0.93x = 5115$

$\dfrac{0.93x}{0.93} = \dfrac{5115}{0.93}$

$x = 5500$ people

63. x = no. of months when costs are equal

$2.65x = 25.44$

$x = \dfrac{25.44}{2.65}$

$x = 9.6$ months

65. <u>Arithmetic Solution</u>

5

$5 + 7 = 12$

$3 \cdot 12 = 36$

$36 - 9 = 27$

$2 \cdot 27 = 54$

$\dfrac{54}{6} = 9$

$9 - 5 = 4$

<u>Algebraic Solution</u>

x

$x + 7$

$3(x + 7) = 3x + 21$

$3x + 21 - 9 = 3x + 12$

$2(3x + 12) = 6x + 24$

$\dfrac{6x + 24}{6} = \dfrac{6x}{6} + \dfrac{24}{6} = x + 4$

$x + 4 - x = 4$

Problem Set 2.7, pp. 100-101

1.   x = first number

2x = second number

x + 2x = 33

3x = 33

x = 11

2x = 22

3.   x = first integer

x + 2 = second integer

x + x + 2 = 54

2x + 2 = 54

2x = 52

x = 26

x + 2 = 28

5.   x = smaller integer

x + 1 = larger integer

3x = 9 + x + 1

3x = 10 + x

2x = 10

x = 5

x + 1 = 6

7.   x = first number

$\frac{3}{5}$x = second number

$x - \frac{3}{5}x = \frac{7}{20}$

$20 \cdot x - 20 \cdot \frac{3}{5}x = 20 \cdot \frac{7}{20}$

20x - 12x = 7

8x = 7

$x = \frac{7}{8}$

$\frac{3}{5}x = \frac{3}{5} \cdot \frac{7}{8} = \frac{21}{40}$

9.

	Age now	Age in 5 yr
Stacy	x	x + 5
Sean	x + 4	x + 9

x + 5 + x + 9 = 38

2x + 14 = 38

2x = 24

x = 12 yr

x + 4 = 16 yr

11.

	Age now	Age in 2 yr
Lisa	x	x + 2
Nancy	x + 9	x + 11

x + 11 = 4(x + 2)

x + 11 = 4x + 8

-3x = -3

x = 1 yr

x + 9 = 10 yr

13.

	Age now	Age 7 yr ago
Brother	x	x - 7
Steve	2x	2x - 7

$2x - 7 = 3(x - 7)$

$2x - 7 = 3x - 21$

$-x = -14$

$x = 14$ yr

$2x = 28$ yr

15.

	Age now	Age in 8 yr	Age 16 yr ago
Mother	x		x - 16
Chris	$\frac{1}{3}$x	$\frac{1}{3}$x + 8	

$\frac{1}{3}x + 8 = x - 16$

$3 \cdot \frac{1}{3}x + 3 \cdot 8 = 3 \cdot x - 3 \cdot 16$

$x + 24 = 3x - 48$

$-2x = -72$

$x = 36$ yr

$\frac{1}{3}x = 12$ yr

17.   x = width

3x - 5 = length

$2\ell \quad + 2w = P$

$2(3x - 5) + 2x = 54$

$6x - 10 + 2x = 54$

$8x - 10 = 54$

$8x = 64$

$x = 8$ m

$3x - 5 = 19$ m

19.   x = width

2x + 24 = length

$2\ell \quad + 2w = P$

$2(2x + 24) + 2x = 210$

$4x + 48 + 2x = 210$

$6x + 48 = 210$

$6x = 162$

$x = 27$ ft

$2x + 24 = 78$ ft

21.   x = length of one side

$$4s = P$$
$$\downarrow \quad \downarrow$$
$$4x = 392$$

$$x = 98 \text{ ft}$$

23.   x = shortest side

x + 2 = middle side

2x = longest side

$$x + x + 2 + 2x = 86$$

$$4x + 2 = 86$$

$$4x = 84$$

$$x = 21 \text{ ft}$$

$$x + 2 = 23 \text{ ft}$$

$$2x = 42 \text{ ft}$$

25.   x = first angle

x - 12 = second angle

2(x - 12) = third angle

$$\angle A + \quad \angle B + \quad \angle C \qquad = 180°$$
$$\downarrow \qquad \downarrow \qquad \downarrow$$
$$x + x - 12 + 2(x - 12) = 180$$

$$x + x - 12 + 2x - 24 = 180$$

$$4x - 36 = 180$$

$$4x = 216$$

$$x = 54°$$

$$x - 12 = 42°$$

$$2(x - 12) = 84°$$

27.   x = length of first piece

x + 2 = length of second piece

x + 5 = length of third piece

$$x + x + 2 + x + 5 = 16$$

$$3x + 7 = 16$$

$$3x = 9$$

$$x = 3 \text{ ft}$$

$$x + 2 = 5 \text{ ft}$$

$$x + 5 = 8 \text{ ft}$$

29.   x = no. of bundles laid
      on the first day

$$x + (x + 2) + (x + 4) + (x + 6) = 92$$

$$4x + 12 = 92$$

$$4x = 80$$

$$x = 20$$

31.   x = no. of miles

$$25 + 0.18x = 70$$

$$0.18x = 45$$

$$x = 250 \text{ mi}$$

Problem Set 2.8, pp. 107-109

1.

	Number	Value
Nickels	x	5x
Dimes	x + 6	10(x + 6)

$$5x + 10(x + 6) = 360$$
$$5x + 10x + 60 = 360$$
$$15x + 60 = 360$$
$$15x = 300$$
$$x = 20 \text{ nickels}$$
$$x + 6 = 26 \text{ dimes}$$

3.

	Number	Value
Nickels	x	5x
Dimes	x - 3	10(x - 3)
Quarters	2x	25(2x)

$$5x + 10(x - 3) + 25(2x) = 750$$
$$5x + 10x - 30 + 50x = 750$$
$$65x - 30 = 750$$
$$65x = 780$$
$$x = 12$$
$$x - 3 = 9$$
$$2x = 24$$

5.

	Number	Value
Nickels	x	5x
Dimes	40 - x	10(40 - x)

$$5x + 10(40 - x) = 315$$
$$5x + 400 - 10x = 315$$
$$400 - 5x = 315$$
$$-5x = -85$$
$$x = 17 \text{ nickels}$$
$$40 - x = 23 \text{ dimes}$$

7.

	Hours	Earned
$5 job	x	5x
$4.50 job	28 - x	4.5(28 - x)

$$5x + 4.5(28 - x) = 134$$
$$5x + 126 - 4.5x = 134$$
$$126 + 0.5x = 134$$
$$0.5x = 8$$
$$x = 16 \text{ hr}$$
$$28 - x = 12 \text{ hr}$$

9.

	Amount	Interest
7% investment	x	0.07x
11% investment	3x	0.11(3x)

$$0.07x + 0.11(3x) = 120$$

$$0.07x + 0.33x = 120$$

$$0.4x = 120$$

$$x = \$300$$

$$3x = \$900$$

11.

	Amount	Interest
12% investment	x	0.12x
8% investment	x + 1260	0.08(x + 1260)

$$0.12x = 0.08(x + 1260)$$

$$0.12x = 0.08x + 100.8$$

$$0.04x = 100.8$$

$$x = \$2520$$

$$x + 1260 = \$3780$$

13.

	Amount	Interest
$9\frac{1}{2}$% investment	x	0.095x
$14\frac{1}{2}$% investment	12,000 - x	0.145(12,000 - x)

$$0.095x = 0.145(12,000 - x)$$

$$0.095x = 1740 - 0.145x$$

$$0.24x = 1740$$

$$x = \$7250$$

$$12,000 - x = \$4750$$

15.

	Amount	Interest
16% investment	0.05x	0.16(0.05x)
14% investment	0.30x	0.14(0.30x)
8% investment	0.65x	0.08(0.65x)

$$0.16(0.05x) + 0.14(0.30x) + 0.08(0.65x) = 76,500$$
$$0.008x + 0.042x + 0.052x = 76,500$$
$$0.102x = 76,500$$
$$x = \$750,000$$

17.

	g of alloy	g of gold
90% alloy	x	0.90x
30% alloy	40	0.30(40)
50% alloy	x + 40	0.50(x + 40)

$$0.90x + 0.30(40) = 0.50(x + 40)$$
$$0.9x + 12 = 0.5x + 20$$
$$0.4x = 8$$
$$x = 20 \text{ g}$$

19.

	kg of solution	kg of acid
60% solution	x	0.60x
45% solution	20	0.45(20)
55% solution	x + 20	0.55(x + 20)

$$0.60x + 0.45(20) = 0.55(x + 20)$$
$$0.6x + 9 = 0.55x + 11$$
$$0.05x = 2$$
$$x = 40 \text{ kg}$$

21.

	Pounds of solution	Pounds of antifreeze
45% solution	60	0.45(60)
100% solution	x	1.00x
50% solution	x + 60	0.50(x + 60)

$$0.45(60) + 1.00x = 0.50(x + 60)$$

$$27 + x = 0.5x + 30$$

$$0.5x = 3$$

$$x = 6 \text{ lb}$$

23.

	Tons of feed	Value of feed
$45.50 feed	x	45.50x
$38 feed	21 - x	38(21 - x)
$40 feed	21	40(21)

$$45.50x + 38(21 - x) = 40(21)$$

$$45.5x + 798 - 38x = 840$$

$$7.5x + 798 = 840$$

$$7.5x = 42$$

$$x = 5.6 \text{ ton}$$

$$21 - x = 15.4 \text{ ton}$$

25.

	r	t	d
2nd ship	35	x	35x
1st ship	25	x + 1	25(x + 1)

$35x = 25(x + 1)$

$35x = 25x + 25$

$10x = 25$

$x = 2.5 \text{ hr}$

27.

	r	t	d
1st cyclist	10	x	10x
2nd cyclist	15	x	15x

$10x + 15x = 75$

$25x = 75$

$x = 3 \text{ hr}$

29.

	r	t	d
Slow train	x	4	4x
Fast train	x + 5	4	4(x + 5)

$4x + 4(x + 5) = 300$

$4x + 4x + 20 = 300$

$8x + 20 = 300$

$8x = 280$

$x = 35 \text{ mph}$

$x + 5 = 40 \text{ mph}$

31.

	r	t	d
Slow runner	x	$1\frac{1}{3}$	$x \cdot 1\frac{1}{3}$
Fast runner	x + 2	1	$(x + 2) \cdot 1$

$$x \cdot 1\frac{1}{3} = (x + 2) \cdot 1$$

$$\frac{4}{3}x = x + 2$$

$$\frac{1}{3}x = 2$$

$$x = 6 \text{ mph}$$

$$x + 2 = 8 \text{ mph}$$

33.

	r	t	d
With wind	320	x	320x
Against wind	280	6 - x	280(6 - x)

$$320x = 280(6 - x)$$

$$320x = 1680 - 280x$$

$$600x = 1680$$

$$x = 2.8 \text{ hr}$$

Therefore the distance is d = 320x = 320(2.8) = 896 mi.

35.

	r	t	d
Fast runner	11	x	11x
Slow runner	8	x	8x

$$11x = 8x + 1$$

$$3x = 1$$

$$x = \frac{1}{3} \text{ hr, or 20 min.}$$

Problem Set 2.9, pp. 116-117

1. $x - 3 < 1$

$x - 3 + 3 < 1 + 3$

$x < 4$

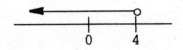

3. $y + 1 \geq -5$

$y + 1 - 1 \geq -5 - 1$

$y \geq -6$

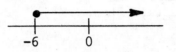

5. $-10 + t > -9$

$-10 + t + 10 > -9 + 10$

$t > 1$

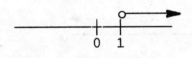

7. $-9 < x - 7$

$-9 + 7 < x - 7 + 7$

$-2 < x$

or

$x > -2$

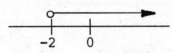

9. $7 \geq p + 7$

$7 - 7 \geq p + 7 - 7$

or

$0 \geq p$

$p \leq 0$

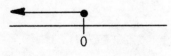

11. $8y + 3 \geq 7y + 11$

$8y + 3 - 7y \geq 7y + 11 - 7y$

$y + 3 \geq 11$

$y + 3 - 3 \geq 11 - 3$

$y \geq 8$

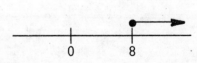

13. $4r \leq 20$

$\dfrac{4r}{4} \leq \dfrac{20}{4}$

$r \leq 5$

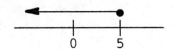

15. $-5t > 35$

$\dfrac{-5t}{-5} < \dfrac{35}{-5}$

$t < -7$

17. $-3p < -6$

$\dfrac{-3p}{-3} > \dfrac{-6}{-3}$

$p > 2$

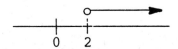

19. $6x > 0$

$\dfrac{6x}{6} > \dfrac{0}{6}$

$x > 0$

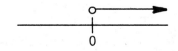

21. $-y \geq 2$

$(-1)(-y) \leq (-1)2$

$y \leq -2$

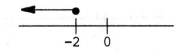

23. $\dfrac{2}{3}x < 4$

$\dfrac{3}{2} \cdot \dfrac{2}{3}x < \dfrac{3}{2} \cdot 4$

$x < 6$

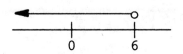

25. $-\frac{4}{5}m \leq 20$

$-\frac{5}{4}(-\frac{4}{5}m) \geq -\frac{5}{4}(20)$

$m \geq -25$

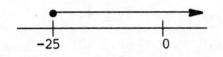

27. $-3z \leq \frac{1}{3}$

$-\frac{1}{3}(-3z) \geq -\frac{1}{3}(\frac{1}{3})$

$z \geq -\frac{1}{9}$

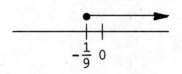

29. $2x - 3 < 7$

$2x - 3 + 3 < 7 + 3$

$2x < 10$

$\frac{2x}{2} < \frac{10}{2}$

$x < 5$

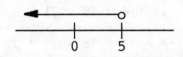

31. $-2p - 5 \geq 3$

$-2p - 5 + 5 \geq 3 + 5$

$-2p \geq 8$

$\frac{-2p}{-2} \leq \frac{8}{-2}$

$p \leq -4$

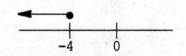

33. $7x - 4 > 4x + 17$

$7x - 4 - 4x > 4x + 17 - 4x$

$3x - 4 > 17$

$3x - 4 + 4 > 17 + 4$

$3x > 21$

$\frac{3x}{3} > \frac{21}{3}$

$x > 7$

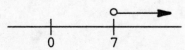

35. $3y + 9 \leq y + 9$

$3y + 9 - y \leq y + 9 - y$

$2y + 9 \leq 9$

$2y + 9 - 9 \leq 9 - 9$

$2y \leq 0$

$\frac{2y}{2} \leq \frac{0}{2}$

$y \leq 0$

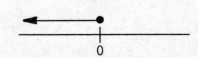

37.
$$21 - 3z > 7z - 19$$
$$21 - 3z - 7z > 7z - 19 - 7z$$
$$21 - 10z > -19$$
$$21 - 10z - 21 > -19 - 21$$
$$-10z > -40$$
$$\frac{-10z}{-10} < \frac{-40}{-10}$$
$$z < 4$$

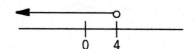

39.
$$8r - 25 \leq 4r + 23$$
$$8r - 25 - 4r \leq 4r + 23 - 4r$$
$$4r - 25 \leq 23$$
$$4r - 25 + 25 \leq 23 + 25$$
$$4r \leq 48$$
$$\frac{4r}{4} \leq \frac{48}{4}$$
$$r \leq 12$$

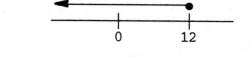

41.
$$-8p < 32 - (7p + 33)$$
$$-8p < 32 - 7p - 33$$
$$-8p < -7p - 1$$
$$-8p + 7p < -7p - 1 + 7p$$
$$-p < -1$$
$$(-1)(-p) > (-1)(-1)$$
$$p > 1$$

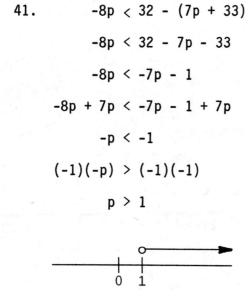

43.
$$4(t + 1) < 12 - 3(t - 2)$$
$$4t + 4 < 12 - 3t + 6$$
$$4t + 4 < 18 - 3t$$
$$4t + 4 + 3t < 18 - 3t + 3t$$
$$7t + 4 < 18$$
$$7t + 4 - 4 < 18 - 4$$
$$7t < 14$$
$$\frac{7t}{7} < \frac{14}{7}$$
$$t < 2$$

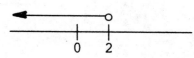

45. $3p - (5p + 15) < p$

$3p - 5p - 15 < p$

$-2p - 15 < p$

$-2p - 15 + 2p < p + 2p$

$-15 < 3p$

$\dfrac{-15}{3} < \dfrac{3p}{3}$

$-5 < p$

or

$p > -5$

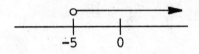

47. $m - (11m + 10) > 0$

$m - 11m - 10 > 0$

$-10m - 10 > 0$

$-10m - 10 + 10m > 0 + 10m$

$-10 > 10m$

$\dfrac{-10}{10} > \dfrac{10m}{10}$

$-1 > m$

or

$m < -1$

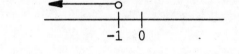

49. $8 - 3(t + 2) > 6t + 5$

$8 - 3t - 6 > 6t + 5$

$2 - 3t > 6t + 5$

$2 - 3t - 6t > 6t + 5 - 6t$

$2 - 9t > 5$

$2 - 9t - 2 > 5 - 2$

$-9t > 3$

$\dfrac{-9t}{-9} < \dfrac{3}{-9}$

$t < -\dfrac{1}{3}$

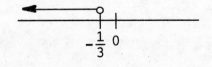

51. $4(2x - 1) > 5 - 2(x - 8)$

$8x - 4 > 5 - 2x + 16$

$8x - 4 > 21 - 2x$

$8x - 4 + 2x > 21 - 2x + 2x$

$10x - 4 > 21$

$10x - 4 + 4 > 21 + 4$

$10x > 25$

$\dfrac{10x}{10} > \dfrac{25}{10}$

$x > \dfrac{5}{2}$

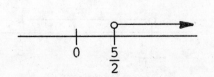

53. $2 < x - 1 < 4$

$2 + 1 < x - 1 + 1 < 4 + 1$

$3 < x < 5$

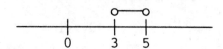

55. $-6 \leq 3y \leq 12$

$\dfrac{-6}{3} \leq \dfrac{3y}{3} \leq \dfrac{12}{3}$

$-2 \leq y \leq 4$

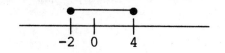

57. $5 < -5y \leq 15$

$\dfrac{5}{-5} > \dfrac{-5y}{-5} \geq \dfrac{15}{-5}$

$-1 > y \geq -3$

or

$-3 \leq y < -1$

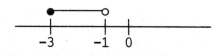

59. $1 < 2r - 9 < 3$

$1 + 9 < 2r - 9 + 9 < 3 + 9$

$10 < 2r < 12$

$\dfrac{10}{2} < \dfrac{2r}{2} < \dfrac{12}{2}$

$5 < r < 6$

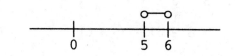

61. $-7 \leq 5 - 4z < 20$

$-7 - 5 \leq 5 - 4z - 5 < 20 - 5$

$-12 \leq -4z < 15$

$\dfrac{-12}{-4} \geq \dfrac{-4z}{-4} > \dfrac{15}{-4}$

$3 \geq z > -\dfrac{15}{4}$

or

$-\dfrac{15}{4} < z \leq 3$

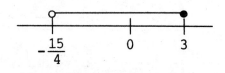

63. $-5 < 6t + 13 < 13$

$-5 - 13 < 6t + 13 - 13 < 13 - 13$

$-18 < 6t < 0$

$\dfrac{-18}{6} < \dfrac{6t}{6} < \dfrac{0}{6}$

$-3 < t < 0$

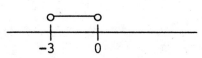

65. $$-14 \le 1 + \frac{5}{3}p \le -9$$

$$-14 - 1 \le 1 + \frac{5}{3}p - 1 \le -9 - 1$$

$$-15 \le \frac{5}{3}p \le -10$$

$$\frac{3}{5}(-15) \le \frac{3}{5}\left(\frac{5}{3}p\right) \le \frac{3}{5}(-10)$$

$$-9 \le p \le -6$$

67. $$-2 < 4 - \frac{1}{2}m < 8$$

$$-2 - 4 < 4 - \frac{1}{2}m - 4 < 8 - 4$$

$$-6 < -\frac{1}{2}m < 4$$

$$-2(-6) > -2\left(-\frac{1}{2}m\right) > -2(4)$$

$$12 > m > -8$$

or

$$-8 < m < 12$$

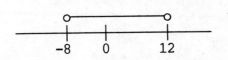

69. $$x + 1 \le 2(x + 3) \le x + 8$$

$$x + 1 \le 2x + 6 \le x + 8$$

$$x + 1 - x \le 2x + 6 - x \le x + 8 - x$$

$$1 \le x + 6 \le 8$$

$$1 - 6 \le x + 6 - 6 \le 8 - 6$$

$$-5 \le x \le 2$$

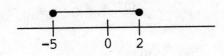

71. x = the number

$$8 - 3x > 17$$

$$8 - 3x - 8 > 17 - 8$$

$$-3x > 9$$

$$\frac{-3x}{-3} < \frac{9}{-3}$$

$$x < -3$$

73.　x = the number

$$6 < 3(x - 5) < 9$$

$$6 < 3x - 15 < 9$$

$$6 + 15 < 3x - 15 + 15 < 9 + 15$$

$$21 < 3x < 24$$

$$\frac{21}{3} < \frac{3x}{3} < \frac{24}{3}$$

$$7 < x < 8$$

75.　　x = width

$$1.5x = \text{length}$$

$$\text{Perimeter} \geq 62.5$$

$$2(1.5x) + 2x \geq 62.5$$

$$3x + 2x \geq 62.5$$

$$5x \geq 62.5$$

$$\frac{5x}{5} \geq \frac{62.5}{5}$$

$$x \geq 12.5 \text{ ft}$$

77.　x = weekly sales

$$\text{Salary under Plan A} > \text{Salary under Plan B}$$

$$0.25x > 350 + 0.05x$$

$$0.25x - 0.05x > 350 + 0.05x - 0.05x$$

$$0.2x > 350$$

$$\frac{0.2x}{0.2} > \frac{350}{0.2}$$

$$x > \$1750$$

79.　x = score on fifth test

$$70 \leq \text{Final average} < 80$$

$$70 \leq \frac{74 + 81 + 68 + 78 + x}{5} < 80$$

$$70 \leq \frac{301 + x}{5} < 80$$

$$5 \cdot 70 \leq 5 \cdot \frac{301 + x}{5} < 5 \cdot 80$$

$$350 \leq 301 + x < 400$$

$$49 \leq x < 99$$

CHAPTER 3

EXPONENTS AND POLYNOMIALS

Problem Set 3.1, pp. 130-132

1. Base 5, exponent 4

3. Base 3, exponent 2

5. Base x, exponent 3

7. $9 \cdot 9 = 9^2$

9. $(-4x)(-4x)(-4x) = (-4x)^3$

11. $-t \cdot t \cdot t \cdot t \cdot t \cdot t = -t^6$

13. $(a^2)(a^2)(a^2) = (a^2)^3$

15. $\frac{a}{b} \cdot \frac{a}{b} \cdot \frac{a}{b} = (\frac{a}{b})^3$

17. $a^3 \cdot a^4 = a^{3+4} = a^7$

19. $3^2 \cdot 3^5 = 3^{2+5} = 3^7$

21. $y^6 \cdot y^4 \cdot y = y^{6+4+1} = y^{11}$

23. $(3x^4)(2x^5) = (3 \cdot 2)(x^4 x^5)$

$$= 6x^{4+5}$$

$$= 6x^9$$

25. $(-5y^6)(4y^3) = (-5 \cdot 4)(y^6 y^3)$

$$= -20y^{6+3}$$

$$= -20y^9$$

27. $(-2p^2)(-p^9) = (-2)(-1)p^2 p^9$

$$= 2p^{2+9}$$

$$= 2p^{11}$$

29. $\frac{a^4}{a^2} = a^{4-2} = a^2$

31. $\frac{9^{10}}{9^2} = 9^{10-2} = 9^8$

33. $\frac{18x^8}{3x} = \frac{18}{3} \cdot \frac{x^8}{x} = 6x^{8-1} = 6x^7$

35. $\frac{-16m^9}{-8m^3} = \frac{-16}{-8} \cdot \frac{m^9}{m^3} = 2m^{9-3} = 2m^6$

37. $\frac{-36x^3 y^{20}}{9xy^4} = \frac{-36}{9} \cdot \frac{x^3}{x} \cdot \frac{y^{20}}{y^4}$

$$= -4x^{3-1} y^{20-4}$$

$$= -4x^2 y^{16}$$

39. $\frac{a^3}{a^3} = a^{3-3} = a^0 = 1$

41. $6^0 = 1$

43. $(-6)^0 = 1$

45. $\left(\frac{3}{4}\right)^0 = 1$

47. $-5x^0 = -5 \cdot 1 = -5$

49. $4^{-2} = \frac{1}{4^2} = \frac{1}{16}$

51. $-11^{-2} = -\frac{1}{11^2} = -\frac{1}{121}$

53. $3 \cdot 2^{-5} = 3 \cdot \frac{1}{2^5} = 3 \cdot \frac{1}{32} = \frac{3}{32}$

55. $(-2)^{-5} = \frac{1}{(-2)^5} = \frac{1}{-32} = -\frac{1}{32}$

57. $\left(\frac{2}{5}\right)^{-2} = \left(\frac{5}{2}\right)^2 = \frac{25}{4}$

59. $\left(\frac{1}{2}\right)^{-3} = \left(\frac{2}{1}\right)^3 = \frac{8}{1} = 8$

61. $\frac{1}{9^{-2}} = 9^2 = 81$

63. $\frac{6}{(-3)^{-4}} = 6 \cdot \frac{1}{(-3)^{-4}} = 6 \cdot (-3)^4 = 6 \cdot 81 = 486$

65. $2^{-1} + 4^{-1} = \frac{1}{2} + \frac{1}{4} = \frac{2}{4} + \frac{1}{4} = \frac{3}{4}$

67. $2 \cdot 3^{-1} + 4 \cdot 6^0 = 2 \cdot \frac{1}{3} + 4 \cdot 1 = \frac{2}{3} + 4 = \frac{2}{3} + \frac{12}{3} = \frac{14}{3}$

69. $9^{-3} \cdot 9^7 = 9^{-3+7} = 9^4$

71. $5^{-2} \cdot 5^{-4} = 5^{-2+(-4)} = 5^{-6} = \frac{1}{5^6}$

73. $x \cdot x^3 \cdot x^{-6} = x^{1+3+(-6)} = x^{-2} = \frac{1}{x^2}$

75. $(3m^{-5})(-5m^9) = 3(-5)m^{-5}m^9 = -15m^{-5+9} = -15m^4$

77. $(-4k^{-3})(-2k^{-1}) = -4(-2)k^{-3}k^{-1} = 8k^{-3+(-1)} = 8k^{-4} = 8 \cdot \frac{1}{k^4} = \frac{8}{k^4}$

79. $\frac{6^4}{6^7} = 6^{4-7} = 6^{-3} = \frac{1}{6^3}$

81. $\frac{r^{-1}}{r} = \frac{r^{-1}}{r^1} = r^{-1-1} = r^{-2} = \frac{1}{r^2}$

83. $\frac{10^{-2}}{10^{-8}} = 10^{-2-(-8)} = 10^{-2+8} = 10^6$

85. $\frac{a^{-3}b^9}{a^4b^{-7}} = a^{-3-4}b^{9-(-7)} = a^{-7}b^{16} = \frac{1}{a^7} \cdot b^{16} = \frac{b^{16}}{a^7}$

87. $\dfrac{4x}{-12x^{-5}y^{-1}} = \dfrac{4}{-12} \cdot \dfrac{x}{x^{-5}} \cdot \dfrac{1}{y^{-1}} = -\dfrac{1}{3} \cdot x^{1-(-5)} \cdot y^{1} = -\dfrac{x^{6}y}{3}$

89. $x^{n} \cdot x = x^{n+1}$

91. $x^{3n+1} \cdot x^{-1} = x^{3n+1+(-1)} = x^{3n}$

93. $\dfrac{x^{n}}{x^{2}} = x^{n-2}$

95. $\dfrac{z^{5m}}{z^{-m}} = z^{5m-(-m)} = z^{5m+m} = z^{6m}$

97. $Q = A \cdot 2^{-t/1600}$

$Q = 100 \cdot 2^{-t/1600}$

(a) $Q = 100 \cdot 2^{-0/1600} = 100 \cdot 2^{0} = 100 \cdot 1 = 100$ g

(b) $Q = 100 \cdot 2^{-1600/1600} = 100 \cdot 2^{-1} = 100 \cdot \dfrac{1}{2} = 50$ g

(c) $Q = 100 \cdot 2^{-3200/1600} = 100 \cdot 2^{-2} = 100 \cdot \dfrac{1}{2^{2}} = 25$ g

99. $\boxed{\text{Clear}}$ 4 $\boxed{y^{x}}$ 2 $\boxed{+/-}$ $\boxed{=}$ 0.0625

$4^{-2} = 0.0625$

101. $\boxed{\text{Clear}}$ 747 $\boxed{y^{x}}$ 0 $\boxed{=}$ 1

$747^{0} = 1$

Problem Set 3.2, pp. 136-138

1. $(x^{3})^{2} = x^{3 \cdot 2} = x^{6}$

3. $(5^{-1})^{-8} = 5^{-1(-8)} = 5^{8}$

5. $(6^{2})^{-5} = 6^{2(-5)} = 6^{-10} = \dfrac{1}{6^{10}}$

7. $(8x)^{2} = 8^{2}x^{2} = 64x^{2}$

9. $(-3pq)^{3} = (-3)^{3}p^{3}q^{3}$

$= -27p^{3}q^{3}$

11. $5(2rs)^{4} = 5 \cdot 2^{4}r^{4}s^{4}$

$= 5 \cdot 16r^{4}s^{4}$

$= 80r^{4}s^{4}$

13. $(x^2y^6)^2 = (x^3)^2(y^6)^2 = x^6y^{12}$

15. $(p^3q^{-1})^4 = (p^3)^4(q^{-1})^4 = p^{12}q^{-4} = p^{12}\cdot\dfrac{1}{q^4} = \dfrac{p^{12}}{q^4}$

17. $(7t^{-3})^{-2} = 7^{-2}(t^{-3})^{-2} = \dfrac{1}{7^2}t^6 = \dfrac{t^6}{49}$

19. $(-2m^6)^3 = (-2)^3(m^6)^3 = -8m^{18}$

21. $(3x)^{-2} = \dfrac{1}{(3x)^2} = \dfrac{1}{3^2x^2} = \dfrac{1}{9x^2}$

23. $3x^{-2} = 3\cdot\dfrac{1}{x^2} = \dfrac{3}{x^2}$

25. $\left(\dfrac{3}{5}\right)^2 = \dfrac{3^2}{5^2} = \dfrac{9}{25}$

27. $\left(\dfrac{x^3}{y^4}\right)^3 = \dfrac{(x^3)^3}{(y^4)^3} = \dfrac{x^9}{y^{12}}$

29. $\left(\dfrac{a^{-1}}{b^{-1}}\right)^2 = \dfrac{(a^{-1})^2}{(b^{-1})^2} = \dfrac{a^{-2}}{b^{-2}} = a^{-2}\cdot\dfrac{1}{b^{-2}} = \dfrac{1}{a^2}\cdot b^2 = \dfrac{b^2}{a^2}$

31. $\left(\dfrac{r^{-2}}{s^5}\right)^{-3} = \dfrac{(r^{-2})^{-3}}{(s^5)^{-3}} = \dfrac{r^6}{s^{-15}} = r^6\cdot\dfrac{1}{s^{-15}} = r^6s^{15}$

33. $(a^3a^2)^3 = (a^5)^3 = a^{15}$

35. $(x^3)^5(x^2)^2 = x^{15}x^4 = x^{19}$

37. $z(z^3)^2(z^2z)^3 = zz^6(z^3)^3 = z^7z^9 = z^{16}$

39. $\dfrac{3^7(x^2)^2}{3^{-1}x^5} = \dfrac{3^7}{3^{-1}}\cdot\dfrac{x^4}{x^5} = 3^{7-(-1)}x^{4-5} = 3^8x^{-1} = 3^8\cdot\dfrac{1}{x} = \dfrac{3^8}{x}$

41. $\dfrac{y^2\cdot y^8\cdot y^5}{(y^3)^2} = \dfrac{y^{15}}{y^6} = y^{15-6} = y^9$

43. $\dfrac{t^{-4} \cdot t}{t^3 \cdot t^{-1}} = \dfrac{t^{-4+1}}{t^{3+(-1)}} = \dfrac{t^{-3}}{t^2} = t^{-3-2} = t^{-5} = \dfrac{1}{t^5}$

45. $\dfrac{m^{11}(m^{-3})^4}{(2m^{-4})^2} = \dfrac{m^{11}m^{-12}}{2^2(m^{-4})^2} = \dfrac{m^{11+(-12)}}{4m^{-8}} = \dfrac{m^{-1}}{4m^{-8}} = \dfrac{1}{4} \cdot \dfrac{m^{-1}}{m^{-8}}$

$$= \dfrac{1}{4} \cdot m^{-1-(-8)}$$

$$= \dfrac{1}{4}m^7$$

$$= \dfrac{m^7}{4}$$

47. $\dfrac{(2a^{-2})^5}{a^{-8} \cdot a^0} = \dfrac{2^5(a^{-2})^5}{a^{-8} \cdot 1} = \dfrac{32a^{-10}}{a^{-8}} = 32a^{-10-(-8)} = 32a^{-2} = 32 \cdot \dfrac{1}{a^2} = \dfrac{32}{a^2}$

49. $\dfrac{(r^{-1}s)^2(rs^{-3})^{-4}}{r^2 s(rs)^{-1}} = \dfrac{(r^{-1})^2 s^2 r^{-4}(s^{-3})^{-4}}{r^2 s r^{-1} s^{-1}} = \dfrac{r^{-2}s^2 r^{-4}s^{12}}{rs^0} = \dfrac{r^{-6}s^{14}}{r}$

$$= r^{-6-1}s^{14}$$

$$= r^{-7}s^{14}$$

$$= \dfrac{s^{14}}{r^7}$$

51. $\left(\dfrac{a^4 b^{-2}}{c^{-3}}\right)^{-2} = \dfrac{(a^4)^{-2}(b^{-2})^{-2}}{(c^{-3})^{-2}} = \dfrac{a^{-8}b^4}{c^6} = a^{-8} \cdot \dfrac{b^4}{c^6} = \dfrac{1}{a^8} \cdot \dfrac{b^4}{c^6} = \dfrac{b^4}{a^8 c^6}$

53. $\dfrac{2a^{-1}}{3b^{-2}} = \dfrac{2}{3} \cdot a^{-1} \cdot \dfrac{1}{b^{-2}} = \dfrac{2}{3} \cdot \dfrac{1}{a} \cdot b^2 = \dfrac{2b^2}{3a}$

55. $\dfrac{(2x^3y^2)^{-2}(2x^{-1}y^4)^3}{(2^{-1}x^2y^{-3})^{-1}} = \dfrac{2^{-2}(x^3)^{-2}(y^2)^{-2}2^3(x^{-1})^3(y^4)^3}{(2^{-1})^{-1}(x^2)^{-1}(y^{-3})^{-1}}$

$$= \dfrac{2^{-2+3}x^{-6}y^{-4}x^{-3}y^{12}}{2x^{-2}y^3}$$

$$= \dfrac{2x^{-9}y^8}{2x^{-2}y^3}$$

$$= x^{-9-(-2)}y^{8-3}$$

$$= x^{-7}y^5$$

$$= \dfrac{y^5}{x^7}$$

57. $(-4x^{-2}y^2)^2 = (-4)^2(x^{-2})^2(y^2)^2 = 16x^{-4}y^4 = 16\cdot\dfrac{1}{x^4}\cdot y^4 = \dfrac{16y^4}{x^4}$

59. $(2x^2)^3(5x^4)^2 = 2^3(x^2)^3 5^2(x^4)^2 = 8x^6 25x^8 = 200x^{14}$

61. $3a^2b^5(4a^3b)^2 = 3a^2b^5 4^2(a^3)^2b^2 = 3a^2b^5 16a^6b^2 = 48a^8b^7$

63. $5r^3s^6(-2rs^3)^3 = 5r^3s^6(-2)^3r^3(s^3)^3 = 5r^3s^6(-8)r^3s^9 = -40r^6s^{15}$

65. $(3p^{-2}q^{-5})^{-2}(3pq^{-2})^5 = 3^{-2}(p^{-2})^{-2}(q^{-5})^{-2}3^5p^5(q^{-2})^5 = 3^{-2}p^4q^{10}3^5p^5q^{-10}$

$$= 3^3p^9q^0$$

$$= 27p^9$$

67. $-m^2(-m^{-1}m^5)^3(-m^{-2}m^{-3})^2 = -m^2(-1)^3(m^{-1})^3(m^5)^3(-1)^2(m^{-2})^2(m^{-3})^2$

$$= -m^2(-1)m^{-3}m^{15}1m^{-4}m^{-6}$$

$$= -(-1)m^{2+(-3)+15+(-4)+(-6)}$$

$$= m^4$$

69. $x^{3n} \cdot x^{4n} \cdot x^2 = x^{3n+4n+2} = x^{7n+2}$

71. $\dfrac{y^{6n}}{y^{-n}} = y^{6n-(-n)} = y^{6n+n} = y^{7n}$

73. $(z^{-8m} \cdot z^{5m})^{-2} = (z^{-8m+5m})^{-2} = (z^{-3m})^{-2} = z^{6m}$

75. $(\dfrac{r^{-2}}{s^4})^{-m} = \dfrac{(r^{-2})^{-m}}{(s^4)^{-m}} = \dfrac{r^{2m}}{s^{-4m}} = r^{2m} \cdot \dfrac{1}{s^{-4m}} = r^{2m}s^{4m}$

77. $(x^{-n}y^{2n})^3 = (x^{-n})^3(y^{2n})^3 = x^{-3n}y^{6n} = \dfrac{1}{x^{3n}} \cdot y^{6n} = \dfrac{y^{6n}}{x^{3n}}$

79. $x^2(x^{m+1})^2 = x^2 x^{2(m+1)} = x^2 x^{2m+2} = x^{2+2m+2} = x^{2m+4}$

81. $4^7 = (2^2)^7 = 2^{14}$

83. $V = p\left[\dfrac{(1 + r)^t - 1}{r}\right]$

$V = 900\left[\dfrac{(1 + 0.10)^{50} - 1}{0.10}\right] = 900\left[\dfrac{(1.1)^{50} - 1}{0.1}\right]$

| Clear | 1.1 | y^x | 50 | − | 1 | = | ÷ | 0.1 | × | 900 | = | 1,047,517.68 |

The value of the fund is approximately \$1,047,518.

Problem Set 3.3, pp. 142-144

1. $82{,}000{,}000 = 8.2 \times 10^7$

3. $3{,}030{,}000{,}000 = 3.03 \times 10^9$

5. $912 = 9.12 \times 10^2$

7. $100{,}000 = 1 \times 10^5 = 10^5$

9. $0.0005 = 5 \times 10^{-4}$

11. $0.74 = 7.4 \times 10^{-1}$

13. $0.000000618 = 6.18 \times 10^{-7}$

15. $52.8 \times 10^3 = 5.28 \times 10 \times 10^3 = 5.28 \times 10^4$

17. $0.93 \times 10^{-4} = 9.3 \times 10^{-1} \times 10^{-4} = 9.3 \times 10^{-5}$

19. $7 \times 10^2 = 700$

21. $6.4 \times 10^{-5} = 0.000064$ 23. $3.18 \times 10^0 = 3.18$

25. $5 \times 10^{-4} = 0.0005$ 27. $2.713 \times 10^2 = 271.3$

29. $1 \times 10^{-3} = 0.001$

31. $2.6 \times 10^{-5} = 0.000026$, $1.94 \times 10^{-2} = 0.0194$, 5.3, $1.7 \times 10^4 = 17,000$

33. $(2 \times 10^8)(4 \times 10^{-5}) = (2 \times 4)(10^8 \times 10^{-5}) = 8 \times 10^3$

35. $(4 \times 10^3)(2.1 \times 10^{-8}) = (4 \times 2.1)(10^3 \times 10^{-8}) = 8.4 \times 10^{-5}$

37. $(2.5 \times 10)(5.4 \times 10^7) = (2.5 \times 5.4)(10 \times 10^7) = 13.5 \times 10^8 = 1.35 \times 10 \times 10^8$

$$= 1.35 \times 10^9$$

39. $\dfrac{6 \times 10^7}{2 \times 10^3} = \dfrac{6}{2} \times \dfrac{10^7}{10^3} = 3 \times 10^{7-3} = 3 \times 10^4$

41. $\dfrac{7.8 \times 10^{-1}}{3 \times 10^{-6}} = \dfrac{7.8}{3} \times \dfrac{10^{-1}}{10^{-6}} = 2.6 \times 10^{-1-(-6)} = 2.6 \times 10^5$

43. $\dfrac{3.5 \times 10^3}{5 \times 10^8} = \dfrac{3.5}{5} \times \dfrac{10^3}{10^8} = 0.7 \times 10^{3-8} = 0.7 \times 10^{-5} = 7 \times 10^{-1} \times 10^{-5}$

$$= 7 \times 10^{-6}$$

45. $(3,000,000)(0.0002) = (3 \times 10^6)(2 \times 10^{-4}) = (3 \times 2)(10^6 \times 10^{-4}) = 6 \times 10^2$

47. $\dfrac{0.000088}{200,000} = \dfrac{8.8 \times 10^{-5}}{2 \times 10^5} = \dfrac{8.8}{2} \times \dfrac{10^{-5}}{10^5} = 4.4 \times 10^{-5-5} = 4.4 \times 10^{-10}$

49. $\dfrac{(300,000,000)(0.0082)}{0.00006} = \dfrac{(3 \times 10^8)(8.2 \times 10^{-3})}{6 \times 10^{-5}}$

$$= \frac{(3)(8.2)}{6} \times \frac{10^8 10^{-3}}{10^{-5}}$$

$$= 4.1 \times 10^{8+(-3)-(-5)}$$

$$= 4.1 \times 10^{10}$$

51. $\dfrac{(0.0000036)(2400)}{(1,200,000)(0.18)} = \dfrac{(3.6 \times 10^{-6})(2.4 \times 10^{3})}{(1.2 \times 10^{6})(1.8 \times 10^{-1})}$

$$= \frac{(3.6)(2.4)}{(1.2)(1.8)} \times \frac{10^{-6} 10^{3}}{10^{6} 10^{-1}}$$

$$= 4 \times 10^{-6+3-6-(-1)}$$

$$= 4 \times 10^{-8}$$

53. $2.5 \times 10^{13} = 25,000,000,000,000$ mi

55. $0.000075 = 7.5 \times 10^{-5}$ cm

57. $1 \times 10^{15} = 10^{15}$

59. $\dfrac{1}{1,000,000} = 0.000001 = 1 \times 10^{-6}$ sec

61. $t = \dfrac{d}{r} = \dfrac{744,000}{186,000} = \dfrac{7.44 \times 10^5}{1.86 \times 10^5} = \dfrac{7.44}{1.86} = 4$ sec

63. $(4.5 \times 10^5 \frac{calc}{sec})(60 \frac{sec}{min})(60 \frac{min}{hr})(10 \ hr)$ = $(4.5 \times 10^5)(6 \times 10)(6 \times 10)(10)$ calc

$$= (4.5 \times 6 \times 6)(10^5 \times 10 \times 10 \times 10) \ calc$$

$$= 162 \times 10^8 \ calc$$

$$= 1.62 \times 10^2 \times 10^8 \ calc$$

$$= 1.62 \times 10^{10} \ calc$$

65. ☐Clear   150,000   ☐x^2 2.25   10

$(150,000)^2 = 2.25 \times 10^{10}$

67. ☐Clear   0.000005   ☐÷   80,000   ☐=   6.25   -11

$\frac{0.000005}{80,000} = 6.25 \times 10^{-11}$

Problem Set 3.4, pp. 150-152

1. x^5, degree is 5, monomial

3. $3x^2 + 7x + 2$, degree is 2, trinomial

5. $-2x^3y^5$, degree is 8, monomial

7. $m^2 + 9m$, degree is 2, binomial

9. $-t^8 + t^6 + 8t^4 - t^2$, degree is 8, polynomial

11. 4, degree is 0 (since $4 = 4 \cdot 1 = 4x^0$), monomial

13. $2x + 5 = 2(6) + 5 = 12 + 5 = 17$

15. $x^2 - 5x + 3 = (-2)^2 - 5(-2) + 3 = 4 + 10 + 3 = 17$

17. $4y^2 + y = 4(5.5)^2 + 5.5 = 4(30.25) + 5.5 = 121 + 5.5 = 126.5$

19. $8t^2 + 6t + 7 = 8(0)^2 + 6(0) + 7 = 0 + 0 + 7 = 7$

21. $3r^3 - r^2 - 8r - 9 = 3(-1)^3 - (-1)^2 - 8(-1) - 9 = 3(-1) - 1 + 8 - 9$

$$= -3 - 1 + 8 - 9$$

$$= -5$$

23. $-z^2 - z + 20 = -3^2 - 3 + 20 = -9 - 3 + 20 = 8$

25. $5p^3 - p^2 + p - 2 = 5(\frac{3}{5})^3 - (\frac{3}{5})^2 + \frac{3}{5} - 2 = 5(\frac{27}{125}) - \frac{9}{25} + \frac{3}{5} - 2$

$$= \frac{27}{25} - \frac{9}{25} + \frac{15}{25} - \frac{50}{25}$$

$$= -\frac{17}{25}$$

27. $3x^4 + 5x^4 = (3 + 5)x^4 = 8x^4$

29. $6m^3 - 10m^3 = (6 - 10)m^3 = -4m^3$

31. $-y^2 - 8y^2 = -1y^2 - 8y^2 = (-1 - 8)y^2 = -9y^2$

33. $4t^2 + 3t + 7t^2 = 4t^2 + 7t^2 + 3t = 11t^2 + 3t$

35. $-x^3y - 4x^3y + 9x^3y = (-1 - 4 + 9)x^3y = 4x^3y$

37. $7p^2q^3 + 11p^3q^2 - 2p^2q^3 = 7p^2q^3 - 2p^2q^3 + 11p^3q^2 = 5p^2q^3 + 11p^3q^2$

39. (a) $(2x + 5) + (3x + 4)$

       $= 2x + 3x + 5 + 4$

       $= 5x + 9$

    (b) $2x + 5$

     (+) $\underline{3x + 4}$

       $5x + 9$

41. (a) $(4x^2 + 7x - 3) + (2x^2 - 5x + 1)$

       $= 4x^2 + 2x^2 + 7x - 5x - 3 + 1$

       $= 6x^2 + 2x - 2$

    (b) $4x^2 + 7x - 3$

     (+) $\underline{2x^2 - 5x + 1}$

       $6x^2 + 2x - 2$

43. (a) $(3x^2 + x^2y^2 - 4y^2) + (5x^2 - x^2y^2 + 2y^2)$

       $= 3x^2 + 5x^2 + x^2y^2 - x^2y^2 - 4y^2 + 2y^2$

       $= 8x^2 - 2y^2$

    (b) $3x^2 + x^2y^2 - 4y^2$

     (+) $\underline{5x^2 - x^2y^2 + 2y^2}$

       $8x^2 \qquad\qquad - 2y^2$

45. (a) $(6x + 3) - (4x - 1)$ (b) $6x + 3$

    $= 6x + 3 - 4x + 1$ $(-)$ $\underline{4x - 1}$

    $= 2x + 4$ $2x + 4$

47. (a) $(8y^3 - y^2 + 5) - (2y^3 + 6y^2 + y - 5)$ (b) $8y^3 - y^2 \qquad + 5$

    $= 8y^3 - y^2 + 5 - 2y^3 - 6y^2 - y + 5$ $(-)$ $\underline{2y^3 + 6y^2 + y - 5}$

    $= 6y^3 - 7y^2 - y + 10$ $6y^3 - 7y^2 - y + 10$

49. (a) $(x^3 - 8x^2y + 9x) - (2x^3 - 5x^2y - 6x + 1)$ (b) $x^3 - 8x^2y + 9x$

    $= x^3 - 8x^2y + 9x - 2x^3 + 5x^2y + 6x - 1$ $(-)$ $\underline{2x^3 - 5x^2y - 6x + 1}$

    $= -x^3 - 3x^2y + 15x - 1$ $-x^3 - 3x^2y + 15x - 1$

51. $6y^3 - 4y^2 + y - 10$ 53. $a^3 - 8a + 5$

  $(+)$ $\underline{5y^3 - y^2 + 2y + 4}$ $(+)$ $\underline{a^3 \qquad - 1}$

    $11y^3 - 5y^2 + 3y - 6$ $2a^3 - 8a + 4$

55. $3x^2 - 4xy + 2y^2$ 57. $10x^2 + 3x + 2$

  $(+)$ $\underline{7x^2 + 5xy - 11y^2}$ $(-)$ $\underline{6x^2 - 2x + 1}$

    $10x^2 + xy - 9y^2$ $4x^2 + 5x + 1$

59. $a^3 + 3a^2 \qquad - 5$ 61. $8x^2 - xy + 6y^3$

  $(-)$ $\underline{-a^3 \qquad - a + 1}$ $(-)$ $\underline{10x^2 - xy - 6y^3}$

    $2a^3 + 3a^2 + a - 6$ $-2x^2 \qquad + 12y^3$

63. $(8x^2 + 4x - 7) + (12x^2 - 3x + 10) = 8x^2 + 4x - 7 + 12x^2 - 3x + 10$

                                     $= 8x^2 + 12x^2 + 4x - 3x - 7 + 10$

                                     $= 20x^2 + x + 3$

65. $(y^2 - 2y - 6) - (3y^2 - 5y + 1) = y^2 - 2y - 6 - 3y^2 + 5y - 1$

                                  $= -2y^2 + 3y - 7$

67. $(a^3 + 4a^2b - 3b^2) - (5a^3 - a^2b + 2b^2) = a^3 + 4a^2b - 3b^2 - 5a^3 + a^2b - 2b^2$

$$= -4a^3 + 5a^2b - 5b^2$$

69. $(15m^3 + 2m^2 - m + 10) + (m^3 - 7m^2 - 1) = 15m^3 + 2m^2 - m + 10 + m^3 - 7m^2 - 1$

$$= 16m^3 - 5m^2 - m + 9$$

71. $(7r^3 - 2r^2 + r - 4) - (3r^3 - r^2 + 6r - 2) = 7r^3 - 2r^2 + r - 4 - 3r^3 + r^2 - 6r + 2$

$$= 4r^3 - r^2 - 5r - 2$$

73. $(2x^2 + 5x + 7) + (x^2 - 6x - 9) - (x^3 - 3x + 8)$

$= 2x^2 + 5x + 7 + x^2 - 6x - 9 - x^3 + 3x - 8$

$= -x^3 + 3x^2 + 2x - 10$

75. $(3x + 1) + (4x - 5) + (x + 7) = 3x + 4x + x + 1 - 5 + 7 = 8x + 3$

77. (a) $0.5n^2 - 1.5n = 0.5(4)^2 - 1.5(4) = 0.5(16) - 6 = 8 - 6 = 2$ diagonals

(b) $0.5n^2 - 1.5n = 0.5(10)^2 - 1.5(10) = 0.5(100) - 15 = 50 - 15 = 35$ diagonals

79. (a) $-16t^2 + 112t + 128 = -16(3)^2 + 112(3) + 128$

$$= -16(9) + 336 + 128$$

$$= -144 + 336 + 128$$

$$= 320 \text{ ft}$$

(b) $-16t^2 + 112t + 128 = -16(8)^2 + 112(8) + 128$

$$= -16(64) + 896 + 128$$

$$= -1024 + 896 + 128$$

$$= 0 \text{ ft}$$

Problem Set 3.5, pp. 154-156

1. $(2x^2)(5x^2) = (2 \cdot 5)(x^2 x^2) = 10x^4$

3. $(-3m^4)(4m^5) = (-3 \cdot 4)(m^4 m^5) = -12m^9$

5. $(9a^3)(5b^2) = (9 \cdot 5)a^3 b^2 = 45a^3 b^2$

7. $(2p)(-7q^2) = 2(-7)pq^2 = -14pq^2$

9. $(-r^3 s)(-8r^2 s^2) = (-1)(-8)r^3 r^2 s s^2 = 8r^5 s^3$

11. $(4xy^5)(-2x^2 y)(3x^3) = 4(-2)3 x x^2 x^3 y^5 y = -24x^6 y^6$

13. $y(y + 5) = y(y) + y(5) = y^2 + 5y$

15. $5x(2x^2 + 6) = 5x(2x^2) + 5x(6) = 10x^3 + 30x$

17. $4r(3 - 2r + r^2) = 4r(3) + 4r(-2r) + 4r(r^2) = 12r - 8r^2 + 4r^3$

19. $-2y^2(4y^2 - 8y - 1) = -2y^2(4y^2) + (-2y^2)(-8y) + (-2y^2)(-1) = -8y^4 + 16y^3 + 2y^2$

21. $-8pq^2(p^2 - 2pq + q^2) = -8pq^2(p^2) + (-8pq^2)(-2pq) + (-8pq^2)(q^2)$

$$= -8p^3 q^2 + 16p^2 q^3 - 8pq^4$$

23. $6t^5(t^3 - 8t^2 + 5t) = 6t^5(t^3) + (6t^5)(-8t^2) + (6t^5)(5t)$

$$= 6t^8 - 48t^7 + 30t^6$$

25. $(x + 2)(x + 4) = x(x + 4) + 2(x + 4) = x^2 + 4x + 2x + 8 = x^2 + 6x + 8$

27. $(t - 3)(t + 3) = t(t + 3) - 3(t + 3) = t^2 + 3t - 3t - 9 = t^2 - 9$

29. $(2x + 5)(x + 3) = 2x(x + 3) + 5(x + 3) = 2x^2 + 6x + 5x + 15 = 2x^2 + 11x + 15$

31. $(4x - 5)(3x + 2) = 4x(3x + 2) - 5(3x + 2) = 12x^2 + 8x - 15x - 10 = 12x^2 - 7x - 10$

33. $(3x - 7y)(2x - y) = 3x(2x - y) - 7y(2x - y) = 6x^2 - 3xy - 14xy + 7y^2$

$$= 6x^2 - 17xy + 7y^2$$

35. $2a(5 + 6a)(3 - 2a) = 2a[5(3 - 2a) + 6a(3 - 2a)]$

$= 2a[15 - 10a + 18a - 12a^2]$

$= 2a[15 + 8a - 12a^2]$

$= 30a + 16a^2 - 24a^3$

37. $(2m + 5)(m^2 - 3m + 4) = 2m(m^2 - 3m + 4) + 5(m^2 - 3m + 4)$

$= 2m^3 - 6m^2 + 8m + 5m^2 - 15m + 20$

$= 2m^3 - m^2 - 7m + 20$

39. $(x - 3)(2x^2 + x - 1) = x(2x^2 + x - 1) - 3(2x^2 + x - 1)$

$= 2x^3 + x^2 - x - 6x^2 - 3x + 3$

$= 2x^3 - 5x^2 - 4x + 3$

41. $(a + 4)(a^3 + 3a - 5) = a(a^3 + 3a - 5) + 4(a^3 + 3a - 5)$

$= a^4 + 3a^2 - 5a + 4a^3 + 12a - 20$

$= a^4 + 4a^3 + 3a^2 + 7a - 20$

43. $(y - 1)(y^3 - 2y^2 + y + 1) = y(y^3 - 2y^2 + y + 1) - 1(y^3 - 2y^2 + y + 1)$

$= y^4 - 2y^3 + y^2 + y - y^3 + 2y^2 - y - 1$

$= y^4 - 3y^3 + 3y^2 - 1$

45. $(4p + 3)(5p^3 - 4p^2 + p - 5) = 4p(5p^3 - 4p^2 + p - 5) + 3(5p^3 - 4p^2 + p - 5)$

$= 20p^4 - 16p^3 + 4p^2 - 20p + 15p^3 - 12p^2 + 3p - 15$

$= 20p^4 - p^3 - 8p^2 - 17p - 15$

47. $(x^2 + 2x - 3)(4x^2 - 5x + 1)$

$= x^2(4x^2 - 5x + 1) + 2x(4x^2 - 5x + 1) - 3(4x^2 - 5x + 1)$

$= 4x^4 - 5x^3 + x^2 + 8x^3 - 10x^2 + 2x - 12x^2 + 15x - 3$

$= 4x^4 + 3x^3 - 21x^2 + 17x - 3$

49.
$$
\begin{array}{r}
p + 5 \\
\underline{p - 8} \\
-8p - 40 \\
\underline{p^2 + 5p} \\
p^2 - 3p - 40
\end{array}
$$

51.
$$
\begin{array}{r}
2r + 5 \\
\underline{2r - 5} \\
-10r - 25 \\
\underline{4r^2 + 10r} \\
4r^2 - 25
\end{array}
$$

53.
$$
\begin{array}{r}
2x^2 + 5x - 4 \\
\underline{x + 3} \\
6x^2 + 15x - 12 \\
\underline{2x^3 + 5x^2 - 4x} \\
2x^3 + 11x^2 + 11x - 12
\end{array}
$$

55.
$$
\begin{array}{r}
8p^2 - p + 6 \\
\underline{2p^2 - 5} \\
-40p^2 + 5p - 30 \\
\underline{16p^4 - 2p^3 + 12p^2} \\
16p^4 - 2p^3 - 28p^2 + 5p - 30
\end{array}
$$

57.
$$
\begin{array}{r}
3m^2 - 4m - 7 \\
\underline{5m^2 + 2m} \\
6m^3 - 8m^2 - 14m \\
\underline{15m^4 - 20m^3 - 35m^2} \\
15m^4 - 14m^3 - 43m^2 - 14m
\end{array}
$$

59.
$$
\begin{array}{r}
x^2 + xy + y^2 \\
\underline{x - y} \\
-x^2y - xy^2 - y^3 \\
\underline{x^3 + x^2y + xy^2} \\
x^3 - y^3
\end{array}
$$

61.
$$
\begin{array}{r}
x + 3 \\
\underline{x + 3} \\
3x + 9 \\
\underline{x^2 + 3x} \\
x^2 + 6x + 9
\end{array}
$$

63.
$$
\begin{array}{r}
4x - 5 \\
\underline{4x - 5} \\
-20x + 25 \\
\underline{16x^2 - 20x} \\
16x^2 - 40x + 25
\end{array}
$$

65.
$$m + 4$$
$$\underline{m + 4}$$
$$4m + 16$$
$$\underline{m^2 + 4m\quad\;}$$
$$m^2 + 8m + 16$$
$$\underline{m + 4}$$
$$4m^2 + 32m + 64$$
$$\underline{m^3 + 8m^2 + 16m\quad\;}$$
$$m^3 + 12m^2 + 48m + 64$$

67.
$$5a - 2b$$
$$\underline{5a - 2b}$$
$$-10ab + 4b^2$$
$$\underline{25a^2 - 10ab\quad\;}$$
$$25a^2 - 20ab + 4b^2$$
$$\underline{5a - 2b}$$
$$-50a^2b + 40ab^2 - 8b^3$$
$$\underline{125a^3 - 100a^2b + 20ab^2\quad\;}$$
$$125a^3 - 150a^2b + 60ab^2 - 8b^3$$

69. $n(n + 2) = n^2 + 2n$

71. $(2x + 3)(7x - 4) = 2x(7x - 4) + 3(7x - 4)$

$$= 14x^2 - 8x + 21x - 12$$

$$= 14x^2 + 13x - 12$$

73. $(10 + 2x)(8 + 2x) - 8 \cdot 10 = 10(8 + 2x) + 2x(8 + 2x) - 80$

$$= 80 + 20x + 16x + 4x^2 - 80$$

$$= 4x^2 + 36x \text{ sq ft}$$

Problem Set 3.6, pp. 160-161

1. $(x + 3)(x + 5) = x^2 + 5x + 3x + 15 = x^2 + 8x + 15$

3. $(x - 2)(x + 4) = x^2 + 4x - 2x - 8 = x^2 + 2x - 8$

5. $(x - 1)(x - 6) = x^2 - 6x - x + 6 = x^2 - 7x + 6$

7. $(3x + 2)(3x - 5) = 9x^2 - 15x + 6x - 10 = 9x^2 - 9x - 10$

9. $(6y - 7)(2y + 3) = 12y^2 + 18y - 14y - 21 = 12y^2 + 4y - 21$

11. $(2r + s)(r + s) = 2r^2 + 2rs + rs + s^2 = 2r^2 + 3rs + s^2$

13. $(4p + q)(p - 7q) = 4p^2 - 28pq + pq - 7q^2 = 4p^2 - 27pq - 7q^2$

15. $(5x - y)(8x - 3y) = 40x^2 - 15xy - 8xy + 3y^2 = 40x^2 - 23xy + 3y^2$

17. $(10a - 3b)(8a + 2b) = 80a^2 + 20ab - 24ab - 6b^2 = 80a^2 - 4ab - 6b^2$

19. $(5 - x)(2 + x) = 10 + 5x - 2x - x^2 = 10 + 3x - x^2$

21. $(m + 5)(m - 5) = m^2 - 5m + 5m - 25 = m^2 - 25$

23. $(2.1x - 1.4)(5.3x + 6.2) = 11.13x^2 + 13.02x - 7.42x - 8.68$

$$= 11.13x^2 + 5.6x - 8.68$$

25. $(t + 4)(t - 4) = t^2 - 4^2 = t^2 - 16$

27. $(p - 8)(p + 8) = p^2 - 8^2 = p^2 - 64$

29. $(6m + 1)(6m - 1) = (6m)^2 - 1^2 = 36m^2 - 1$

31. $(2x - y)(2x + y) = (2x)^2 - y^2 = 4x^2 - y^2$

33. $(3r + 7s)(3r - 7s) = (3r)^2 - (7s)^2 = 9r^2 - 49s^2$

35. $(9x + 3.1)(9x - 3.1) = (9x)^2 - (3.1)^2 = 81x^2 - 9.61$

37. $(x + 2)^2 = x^2 + 2(x)(2) + 2^2 = x^2 + 4x + 4$

39. $(z - 5)^2 = z^2 - 2(z)(5) + 5^2 = z^2 - 10z + 25$

41. $(2y - 3)^2 = (2y)^2 - 2(2y)(3) + 3^2 = 4y^2 - 12y + 9$

43. $(10a + b)^2 = (10a)^2 + 2(10a)(b) + b^2 = 100a^2 + 20ab + b^2$

45. $(5x - 9y)^2 = (5x)^2 - 2(5x)(9y) + (9y)^2 = 25x^2 - 90xy + 81y^2$

47. $(8k + \frac{1}{2})^2 = (8k)^2 + 2(8k)(\frac{1}{2}) + (\frac{1}{2})^2 = 64k^2 + 8k + \frac{1}{4}$

49. $(6y - 5)(3y + 4) = 18y^2 + 24y - 15y - 20 = 18y^2 + 9y - 20$

51. $(1.1p + 2.5)^2 = (1.1p)^2 + 2(1.1p)(2.5) + (2.5)^2 = 1.21p^2 + 5.5p + 6.25$

53. $(3k + \frac{2}{3})(3k - \frac{2}{3}) = (3k)^2 - (\frac{2}{3})^2 = 9k^2 - \frac{4}{9}$

55. $(t^2 - 1)(t^2 - 9) = t^4 - 9t^2 - t^2 + 9 = t^4 - 10t^2 + 9$

57. $(z^2 - 3)(z^2 + 3) = (z^2)^2 - 3^2 = z^4 - 9$

59. $(5 - x^3)^2 = 5^2 - 2(5)(x^3) + (x^3)^2 = 25 - 10x^3 + x^6$

61. $(x + 1)^2 - (x + 6)(x - 3) = x^2 + 2(x)(1) + 1^2 - (x^2 - 3x + 6x - 18)$

$$= x^2 + 2x + 1 - x^2 + 3x - 6x + 18$$

$$= -x + 19$$

63. $(z - 6)(z + 6) - (z - 10)^2 = z^2 - 6^2 - (z^2 - 2(z)(10) + 10^2)$

$$= z^2 - 36 - (z^2 - 20z + 100)$$

$$= z^2 - 36 - z^2 + 20z - 100$$

$$= 20z - 136$$

65. $(x - y)^2 - (x + y)^2 = x^2 - 2xy + y^2 - (x^2 + 2xy + y^2)$

$$= x^2 - 2xy + y^2 - x^2 - 2xy - y^2$$

$$= -4xy$$

Problem Set 3.7, pp. 162-163

1. $\frac{12x^{10}}{3x^5} = \frac{12}{3} \cdot \frac{x^{10}}{x^5} = 4x^{10-5} = 4x^5$

3. $\dfrac{-28y^2}{4y} = \dfrac{-28}{4} \cdot \dfrac{y^2}{y} = -7y^{2-1} = -7y$

5. $\dfrac{10m}{5m^2} = \dfrac{10}{5} \cdot \dfrac{m}{m^2} = 2 \cdot \dfrac{1}{m} = \dfrac{2}{m}$

7. $\dfrac{-8r^6s^3}{8r^3} = \dfrac{-8}{8} \cdot \dfrac{r^6}{r^3} \cdot \dfrac{s^3}{1} = -1r^{6-3}s^3 = -r^3s^3$

9. $\dfrac{20a^4b^2}{2a^2b^4} = \dfrac{20}{2} \cdot \dfrac{a^4}{a^2} \cdot \dfrac{b^2}{b^4} = 10 \cdot a^2 \cdot \dfrac{1}{b^2} = \dfrac{10a^2}{b^2}$

11. $\dfrac{4x^2y^4}{8x^3y} = \dfrac{4}{8} \cdot \dfrac{x^2}{x^3} \cdot \dfrac{y^4}{y} = \dfrac{1}{2} \cdot \dfrac{1}{x} \cdot y^3 = \dfrac{y^3}{2x}$

13. $\dfrac{5x + 5}{5} = \dfrac{5x}{5} + \dfrac{5}{5} = x + 1$

15. $\dfrac{8x + 8y}{8} = \dfrac{8x}{8} + \dfrac{8y}{8} = x + y$

17. $\dfrac{6x - 4}{-2} = \dfrac{6x}{-2} - \dfrac{4}{-2} = -3x - (-2) = -3x + 2$

19. $\dfrac{y^3 - y^2 + y}{y} = \dfrac{y^3}{y} - \dfrac{y^2}{y} + \dfrac{y}{y} = y^2 - y + 1$

21. $\dfrac{-12p^3 + 18p^2 + 6p}{-6p} = \dfrac{-12p^3}{-6p} + \dfrac{18p^2}{-6p} + \dfrac{6p}{-6p} = 2p^2 - 3p - 1$

23. $\dfrac{6xy^2 - 2x^2y + xy}{xy} = \dfrac{6xy^2}{xy} - \dfrac{2x^2y}{xy} + \dfrac{xy}{xy} = 6y - 2x + 1$

25. $\dfrac{45a^5b^3 + 75a^3b^5 - 30a^2b^5}{15a^2b^3} = \dfrac{45a^5b^3}{15a^2b^3} + \dfrac{75a^3b^5}{15a^2b^3} - \dfrac{30a^2b^5}{15a^2b^3}$

$$= 3a^3 + 5ab^2 - 2b^2$$

27. $\dfrac{27m^5 + 2m^3 - 9m^2}{3m^2} = \dfrac{27m^5}{3m^2} + \dfrac{2m^3}{3m^2} - \dfrac{9m^2}{3m^2} = 9m^3 + \dfrac{2}{3}m - 3$

29. $\dfrac{64x^3 + 3x^2 - 4x}{4x^2} = \dfrac{64x^3}{4x^2} + \dfrac{3x^2}{4x^2} - \dfrac{4x}{4x^2} = 16x + \dfrac{3}{4} - \dfrac{1}{x}$

31. $\dfrac{a^2 + 2ab + b^2}{2a} = \dfrac{a^2}{2a} + \dfrac{2ab}{2a} + \dfrac{b^2}{2a} = \dfrac{a}{2} + b + \dfrac{b^2}{2a}$

33. $\dfrac{18k^4 - 6k^3 + 9k^2 - 15k + 2}{3k^2} = \dfrac{18k^4}{3k^2} - \dfrac{6k^3}{3k^2} + \dfrac{9k^2}{3k^2} - \dfrac{15k}{3k^2} + \dfrac{2}{3k^2}$

$$= 6k^2 - 2k + 3 - \dfrac{5}{k} + \dfrac{2}{3k^2}$$

35. $\dfrac{9m^2 + 6m - 12}{3} = \dfrac{9m^2}{3} + \dfrac{6m}{3} - \dfrac{12}{3} = 3m^2 + 2m - 4$

37. $\dfrac{21x^5 - 30x^3 - 3}{3} = \dfrac{21x^5}{3} - \dfrac{30x^3}{3} - \dfrac{3}{3} = 7x^5 - 10x^3 - 1$

39. $\dfrac{3y + 1}{3} = \dfrac{3y}{3} + \dfrac{1}{3} = y + \dfrac{1}{3}$

41. $\dfrac{z + 9}{3} = \dfrac{z}{3} + \dfrac{9}{3} = \dfrac{z}{3} + 3$

43. $\dfrac{20r^5 - 28r^4}{4r^3} = \dfrac{20r^5}{4r^3} - \dfrac{28r^4}{4r^3} = 5r^2 - 7r$

45. $\dfrac{12r^6 + 8r^4 - 4r^3}{4r^3} = \dfrac{12r^6}{4r^3} + \dfrac{8r^4}{4r^3} - \dfrac{4r^3}{4r^3} = 3r^3 + 2r - 1$

47. $\dfrac{4r^8s^4 - 12r^6s^2 - 2r^4s + 8r^2}{4r^3} = \dfrac{4r^8s^4}{4r^3} - \dfrac{12r^6s^2}{4r^3} - \dfrac{2r^4s}{4r^3} + \dfrac{8r^2}{4r^3}$

$$= r^5s^4 - 3r^3s^2 - \dfrac{rs}{2} + \dfrac{2}{r}$$

49. True. $\dfrac{ab}{b} = \dfrac{a \cdot \cancel{b}}{1 \cdot \cancel{b}} = \dfrac{a}{1} = a$

51. False. $\dfrac{x + 5}{5} = \dfrac{x}{5} + \dfrac{5}{5} = \dfrac{x}{5} + 1$

53. Average $= \dfrac{n + (n + 1) + (n + 2)}{3} = \dfrac{3n + 3}{3} = \dfrac{3n}{3} + \dfrac{3}{3} = n + 1$

Problem Set 3.8, p. 168

1.
$$
\begin{array}{r}
x + 7 \\
x + 3 \,\overline{\smash{\big)}\, x^2 + 10x + 21} \\
\underline{x^2 + 3x} \\
7x + 21 \\
\underline{7x + 21} \\
0
\end{array}
$$

3.
$$
\begin{array}{r}
4x + 3 \\
x + 2 \,\overline{\smash{\big)}\, 4x^2 + 11x + 6} \\
\underline{4x^2 + 8x} \\
3x + 6 \\
\underline{3x + 6} \\
0
\end{array}
$$

5.
$$
\begin{array}{r}
3x + 4 \\
3x - 2 \,\overline{\smash{\big)}\, 9x^2 + 6x + 8} \\
\underline{9x^2 - 6x} \\
12x + 8 \\
\underline{12x - 8} \\
16
\end{array}
$$

7.
$$
\begin{array}{r}
5y + 1 \\
5y - 3 \,\overline{\smash{\big)}\, 25y^2 - 10y - 13} \\
\underline{25y^2 - 15y} \\
5y - 13 \\
\underline{5y - 3} \\
-10
\end{array}
$$

9.
$$2p + 4 \overline{\smash{\big)}\,2p^2 + 0p + 1} \atop p - 2$$

$$\underline{2p^2 + 4p}$$

$$-4p + 1$$

$$\underline{-4p - 8}$$

$$9$$

11.
$$4z + 1 \overline{\smash{\big)}\,8z^2 - 2z + 1} \atop 2z - 1$$

$$\underline{8z^2 + 2z}$$

$$-4z + 1$$

$$\underline{-4z - 1}$$

$$2$$

13.
$$2x + 3 \overline{\smash{\big)}\,8x^3 - 4x^2 - 14x + 21} \atop 4x^2 - 8x + 5$$

$$\underline{8x^3 + 12x^2}$$

$$-16x^2 - 14x$$

$$\underline{-16x^2 - 24x}$$

$$10x + 21$$

$$\underline{10x + 15}$$

$$6$$

15.
$$y + 3 \overline{\smash{\big)}\,y^3 + 0y^2 - 8y + 3} \atop y^2 - 3y + 1$$

$$\underline{y^3 + 3y^2}$$

$$-3y^2 - 8y$$

$$\underline{-3y^2 - 9y}$$

$$y + 3$$

$$\underline{y + 3}$$

$$0$$

17.
$$2m - 3 \overline{\smash{\big)}\,2m^3 - 11m^2 + 0m + 27} \atop m^2 - 4m - 6$$

$$\underline{2m^3 - 3m^2}$$

$$-8m^2 + 0m$$

$$\underline{-8m^2 + 12m}$$

$$-12m + 27$$

$$\underline{-12m + 18}$$

$$9$$

19.
$$4m + 3 \overline{\smash{\big)}\,8m^4 - 10m^3 + 0m^2 - 19m - 21} \atop 2m^3 - 4m^2 + 3m - 7$$

$$\underline{8m^4 + 6m^3}$$

$$-16m^3 + 0m^2$$

$$\underline{-16m^3 - 12m^2}$$

$$12m^2 - 19m$$

$$\underline{12m^2 + 9m}$$

$$-28m - 21$$

$$\underline{-28m - 21}$$

$$0$$

21.

$$
\begin{array}{r}
3p^3 + 2p^2 + p + 7 \\
6p - 5 \overline{) 18p^4 - 3p^3 - 4p^2 + 37p - 5}
\end{array}
$$

$$\underline{18p^4 - 15p^3}$$

$$12p^3 - 4p^2$$

$$\underline{12p^3 - 10p^2}$$

$$6p^2 + 37p$$

$$\underline{6p^2 - 5p}$$

$$42p - 5$$

$$\underline{42p - 35}$$

$$30$$

23.

$$
\begin{array}{r}
3x^3 - 2x^2 + 3x - 2 \\
3x + 2 \overline{) 9x^4 + 0x^3 + 5x^2 + 0x - 4}
\end{array}
$$

$$\underline{9x^4 + 6x^3}$$

$$-6x^3 + 5x^2$$

$$\underline{-6x^3 - 4x^2}$$

$$9x^2 + 0x$$

$$\underline{9x^2 + 6x}$$

$$-6x - 4$$

$$\underline{-6x - 4}$$

$$0$$

25.
$$
\begin{array}{r}
x^2 + 5x - 6 \\
x^2 + x - 2 \overline{\smash{\big)}\ x^4 + 6x^3 - 3x^2 - 16x + 12} \\
\underline{x^4 + x^3 - 2x^2} \\
5x^3 - x^2 - 16x \\
\underline{5x^3 + 5x^2 - 10x} \\
-6x^2 - 6x + 12 \\
\underline{-6x^2 - 6x + 12} \\
0
\end{array}
$$

27.
$$
\begin{array}{r}
x^2 + 4x + 1 \\
x^2 + 0x - 1 \overline{\smash{\big)}\ x^4 + 4x^3 + 0x^2 - 7x + 1} \\
\underline{x^4 + 0x^3 - x^2} \\
4x^3 + x^2 - 7x \\
\underline{4x^3 + 0x^2 - 4x} \\
x^2 - 3x + 1 \\
\underline{x^2 + 0x - 1} \\
-3x + 2
\end{array}
$$

29.

$$\begin{array}{r} 2x^3 - 3x^2 - 4x - 1 \\ 2x^2 - x + 3 \overline{\smash{)}\, 4x^5 - 8x^4 + x^3 - 7x^2 - 10x - 3} \end{array}$$

$$\underline{4x^5 - 2x^4 + 6x^3}$$

$$-6x^4 - 5x^3 - 7x^2$$

$$\underline{-6x^4 + 3x^3 - 9x^2}$$

$$-8x^3 + 2x^2 - 10x$$

$$\underline{-8x^3 + 4x^2 - 12x}$$

$$-2x^2 + 2x - 3$$

$$\underline{-2x^2 + x - 3}$$

$$x$$

31.

$$\begin{array}{r} 2x^2 + x - 2 \\ 2x - 1 \overline{\smash{)}\, 4x^3 + 0x^2 - 5x + 0} \end{array}$$

$$\underline{4x^3 - 2x^2}$$

$$2x^2 - 5x$$

$$\underline{2x^2 - x}$$

$$-4x + 0$$

$$\underline{-4x + 2}$$

$$-2$$

33.

$$\begin{array}{r} m^2 - 3m + 9 \\ m + 3 \overline{\smash{)}\, m^3 + 0m^2 + 0m + 27} \end{array}$$

$$\underline{m^3 + 3m^2}$$

$$-3m^2 + 0m$$

$$\underline{-3m^2 - 9m}$$

$$9m + 27$$

$$\underline{9m + 27}$$

$$0$$

35.

$$3y - 6 \overline{\smash{\big)}\ 3y^2 - 4y + 1} \quad \left(y + \frac{2}{3} \right)$$

$$\underline{3y^2 - 6y}$$

$$2y + 1$$

$$\underline{2y - 4}$$

$$5$$

37. True

39. False

NOTES

CHAPTER 4

FACTORING AND QUADRATIC EQUATIONS

Problem Set 4.1, pp. 179-180

1. $12 = 2 \cdot 6 = 2 \cdot 2 \cdot 3 = 2^2 \cdot 3$

3. $126 = 2 \cdot 63 = 2 \cdot 3 \cdot 21 = 2 \cdot 3 \cdot 3 \cdot 7$
   $= 2 \cdot 3^2 \cdot 7$

5. 37 is a prime number.

7. $2520 = 2 \cdot 1260 = 2 \cdot 2 \cdot 630$
   $= 2 \cdot 2 \cdot 2 \cdot 315$
   $= 2 \cdot 2 \cdot 2 \cdot 3 \cdot 105$
   $= 2 \cdot 2 \cdot 2 \cdot 3 \cdot 3 \cdot 35$
   $= 2 \cdot 2 \cdot 2 \cdot 3 \cdot 3 \cdot 5 \cdot 7$
   $= 2^3 \cdot 3^2 \cdot 5 \cdot 7$

9. $4 = 2^2$
   $12 = 2^2 \cdot 3$
   $GCF = 2^2 = 4$

11. $36 = 2^2 \cdot 3^2$
    $90 = 2 \cdot 3^2 \cdot 5$
    $GCF = 2 \cdot 3^2 = 18$

13. $8 = 2^3$
    $10 = 2 \cdot 5$
    $15 = 3 \cdot 5$
    $GCF = 1$

15. $336 = 2^4 \cdot 3 \cdot 7$
    $252 = 2^2 \cdot 3^2 \cdot 7$
    $420 = 2^2 \cdot 3 \cdot 5 \cdot 7$
    $GCF = 2^2 \cdot 3 \cdot 7 = 84$

17. $7p = 7 \cdot p$
    $35 = 5 \cdot 7$
    $GCF = 7$

19. $3x^3 = 3 \cdot x^3$
    $5x^6 = 5 \cdot x^6$
    $GCF = x^3$

21. $-50x^3 = -2 \cdot 5^2 \cdot x^3$

  $75x^5 = 3 \cdot 5^2 \cdot x^5$

  $25x^{15} = 5^2 \cdot x^{15}$

  $GCF = 5^2 \cdot x^3 = 25x^3$

23. $24x^{15}y^8 = 2^3 \cdot 3 \cdot x^{15} \cdot y^8$

  $28x^7y^8 = 2^2 \cdot 7 \cdot x^7 \cdot y^8$

  $36x^{10}y^6 = 2^2 \cdot 3^2 \cdot x^{10} \cdot y^6$

  $GCF = 2^2 \cdot x^7 \cdot y^6 = 4x^7y^6$

25. $x(3a - 4) = x(3a - 4)$

  $y(3a - 4) = y(3a - 4)$

  $GCF = 3a - 4$

27. $m + 5 = m + 5$

  $(m + 5)^2 = (m + 5)^2$

  $GCF = m + 5$

29. $2m + 6 = 2 \cdot m + 2 \cdot 3$

  $= 2(m + 3)$

31. $6ab - 15ac = 3a \cdot 2b - 3a \cdot 5c$

  $= 3a(2b - 5c)$

33. $12x^2 + 20x = 4x \cdot 3x + 4x \cdot 5$

  $= 4x(3x + 5)$

35. $k^3 + k = k \cdot k^2 + k \cdot 1$

  $= k(k^2 + 1)$

37. $36m^{12} - 24m^8 = 12m^8 \cdot 3m^4 - 12m^8 \cdot 2$

  $= 12m^8(3m^4 - 2)$

39. $3p + 8$ is prime.

41. $-x - 2y = (-1)x + (-1)2y$

  $= -1(x + 2y)$

  $= -(x + 2y)$

43. $r + s - t = (-1)(-r) + (-1)(-s) + (-1)t$

  $= (-1)(-r + (-s) + t$

  $= -(-r - s + t)$

45. $-4a^2 - 9a + 10 = (-1)4a^2 + (-1)9a + (-1)(-10) = (-1)(4a^2 + 9a + (-10))$

  $= -(4a^2 + 9a - 10)$

47. $3x^2 + 6x + 9 = 3(x^2 + 2x + 3)$

49. $144r^2 + 125rs - 24s^2$ is prime.

51. $-12y^3 + 18y^2 - 6y = -6y(2y^2 - 3y + 1)$

53. $8z^{75} - 32z^{50} - 16z^{25} = 8z^{25}(z^{50} - 4z^{25} - 2)$

55. $24x^5y^2 - 36x^3y^3 + 48x^2y^5 = 12x^2y^2(2x^3 - 3xy + 4y^3)$

57. $30r^8s^5t^3 + 72r^5s^7t - 18r^6s^9 = 6r^5s^5(5r^3t^3 + 12s^2t - 3rs^4)$

59. $2a^2bc + 5ab^2c + 9abc^2 = abc(2a + 5b + 9c)$

61. $-22m^{53}p + 55m^{37}p^2 - 33m^{24}p^3 + 11m^{17}p^4 = -11m^{17}p(2m^{36} - 5m^{20}p + 3m^7p^2 - p^3)$

63. $(a + 2b)x + (a + 2b)y = (a + 2b)(x + y)$

65. $x(x + 8) - 3(x + 8) = (x + 8)(x - 3)$

67. $x(y^2 + 9) - 2(y^2 + 9) = (y^2 + 9)(x - 2)$

69. $(4r - 5s)x + (4r - 5s) = (4r - 5s)x + (4r - 5s) \cdot 1 = (4r - 5s)(x + 1)$

71. $(p - 7) + (p - 7)^2 = (p - 7) \cdot 1 + (p - 7)^2 = (p - 7)(1 + (p - 7)) = (p - 7)(p - 6)$

73. $r(r + 1) + (r + 1) = r(r + 1) + (r + 1) \cdot 1 = (r + 1)(r + 1) = (r + 1)^2$

75. $\pi r \ell + \pi r^2 = \pi r(\ell + r)$

Problem Set 4.2, p. 183

1. $ax + ay + bx + by = a(x + y) + b(x + y) = (x + y)(a + b)$

3. $xy + 3y + 2x + 6 = y(x + 3) + 2(x + 3) = (x + 3)(y + 2)$

5. $rs - 8r + 2s - 16 = r(s - 8) + 2(s - 8) = (s - 8)(r + 2)$

7. $3x^2 + 3xy + 2x + 2y = 3x(x + y) + 2(x + y) = (x + y)(3x + 2)$

9. $t^3 - 6t^2 + 6t - 36 = t^2(t - 6) + 6(t - 6) = (t - 6)(t^2 + 6)$

11. $x^2 - x + 7x - 7 = x(x - 1) + 7(x - 1) = (x - 1)(x + 7)$

13. $4m^2 + 8m - 3m - 6 = 4m(m + 2) - 3(m + 2) = (m + 2)(4m - 3)$

15. $2y^2 - 9y - 4y + 18 = y(2y - 9) - 2(2y - 9) = (2y - 9)(y - 2)$

17. $10r^2 + 6rs + 5rs + 3s^2 = 2r(5r + 3s) + s(5r + 3s) = (5r + 3s)(2r + s)$

19. $6x^2 - 14ax + 15ax - 35a^2 = 2x(3x - 7a) + 5a(3x - 7a) = (3x - 7a)(2x + 5a)$

21. $k^3 + 5k^2 + k + 5 = k^2(k + 5) + 1(k + 5) = (k + 5)(k^2 + 1)$

23. $t^2 + 2t + t + 2 = t(t + 2) + 1(t + 2) = (t + 2)(t + 1)$

25. $p^2 - 8p - p + 8 = p(p - 8) - 1(p - 8) = (p - 8)(p - 1)$

27. $m^2 + m + m + 1 = m(m + 1) + 1(m + 1) = (m + 1)(m + 1) = (m + 1)^2$

29. $16x^2 - 4x - 4x + 1 = 4x(4x - 1) - 1(4x - 1) = (4x - 1)(4x - 1) = (4x - 1)^2$

31. $x^2y - ax - xy + a = x(xy - a) - 1(xy - a) = (xy - a)(x - 1)$

33. $12p^2 - 4pq - 3pq + q^2 = 4p(3p - q) - q(3p - q) = (3p - q)(4p - q)$

35. $25k^3 - 75k^2 - 6k + 18 = 25k^2(k - 3) - 6(k - 3) = (k - 3)(25k^2 - 6)$

37. $2ax - 2ay + 2bx - 2by = 2(ax - ay + bx - by)$

$$= 2[a(x - y) + b(x - y)]$$

$$= 2(x - y)(a + b)$$

39. $5ab + 20b + 10a + 40 = 5(ab + 4b + 2a + 8)$

$$= 5[b(a + 4) + 2(a + 4)]$$

$$= 5(a + 4)(b + 2)$$

41. $36p^3 - 72p^2 + 48p - 96 = 12(3p^3 - 6p^2 + 4p - 8)$

$$= 12[3p^2(p - 2) + 4(p - 2)]$$

$$= 12(p - 2)(3p^2 + 4)$$

43. $9r^4s + 63r^3s^2 - 2r^3s^2 - 14r^2s^3 = r^2s(9r^2 + 63rs - 2rs - 14s^2)$

$$= r^2s[9r(r + 7s) - 2s(r + 7s)]$$

$$= r^2s(r + 7s)(9r - 2s)$$

45. $4xy^2 - 4y^2 - 4xy + 4y = 4y(xy - y - x + 1)$

$$= 4y[y(x - 1) - 1(x - 1)]$$

$$= 4y(x - 1)(y - 1)$$

47. $x^2 - xy + x + 3x - 3y + 3 = x(x - y + 1) + 3(x - y + 1)$

$$= (x - y + 1)(x + 3)$$

49. $P(1 + r) + P(1 + r)r = P(1 + r) \cdot 1 + P(1 + r) \cdot r$

$$= P(1 + r)(1 + r)$$

$$= P(1 + r)^2$$

Problem Set 4.3, pp. 188-189

1. $x^2 + 8x + 15 = (x + 3)(x + 5)$

3. $x^2 + 4x + 4 = (x + 2)^2$

5. $x^2 - 5x + 6 = (x - 2)(x - 3)$

7. $y^2 - 5y + 4 = (y - 1)(y - 4)$

9. $y^2 - 15y + 54 = (y - 6)(y - 9)$

11. $4m + m^2 - 12 = m^2 + 4m - 12 = (m + 6)(m - 2)$

13. $m^2 - m - 12 = (m - 4)(m + 3)$

15. $p^2 + 5p - 4$ is prime.

17. $z^2 - 23z - 24 = (z - 24)(z + 1)$

19. $x^2 - 15x + 14 = (x - 1)(x - 14)$

21. $t^2 + 14t - 51 = (t - 3)(t + 17)$

23. $y^2 - 8y - 20 = (y - 10)(y + 2)$

25. $x^2 + 12xy + 35y^2 = (x + 5y)(x + 7y)$

27. $m^2 - 10mn + 25n^2 = (m - 5n)^2$

29. $x^2 + 3xy - 28y^2 = (x + 7y)(x - 4y)$

31. $r^2 - 5rs - 6s^2 = (r - 6s)(r + s)$

33. $z^2 + 6zk - 18k^2$ is prime.

35. $p^2 + 36q^2 - 13pq = p^2 - 13pq + 36q^2 = (p - 4q)(p - 9q)$

37. $-x^2 + 11x - 10 = -(x^2 - 11x + 10) = -(x - 1)(x - 10)$

39. $45 + 4z - z^2 = -z^2 + 4z + 45 = -(z^2 - 4z - 45) = -(z - 9)(z + 5)$

41. $-r^2 - 11rs - 30s^2 = -(r^2 + 11rs + 30s^2) = -(r + 5s)(r + 6s)$

43. $4x^2 + 16x + 16 = 4(x^2 + 4x + 4) = 4(x + 2)^2$

45. $a^4 + 2a^3 + a^2 = a^2(a^2 + 2a + 1) = a^2(a + 1)^2$

47. $m^3 + 6m^2 - 27m = m(m^2 + 6m - 27) = m(m + 9)(m - 3)$

49. $5r^4 - 60r^3 + 55r^2 = 5r^2(r^2 - 12r + 11) = 5r^2(r - 1)(r - 11)$

51. $-10p^3q - 50p^2q^2 + 360pq^3 = -10pq(p^2 + 5pq - 36q^2) = -10pq(p + 9q)(p - 4q)$

53. $6a^2x^5 - 48a^2x^4 - 72a^2x^3 = 6a^2x^3(x^2 - 8x - 12)$

55. $t^4 + 5t^2 + 6 = (t^2 + 2)(t^2 + 3)$

57. $k^6 + 3k^3 - 10 = (k^3 + 5)(k^3 - 2)$

59. $x^4 + 4x^2y^2 + 3y^4 = (x^2 + y^2)(x^2 + 3y^2)$

61. $(2x - 3)(x + 5) = 2x^2 + 10x - 3x - 15 = 2x^2 + 7x - 15$

Problem Set 4.4, p. 194

1. $2x^2 + 11x + 5 = (2x + 1)(x + 5)$

3. $2z^2 - 7z + 5 = (2z - 5)(z - 1)$

5. $7p^2 - 16p + 4 = (7p - 2)(p - 2)$

7. $5m^2 + 2m - 3 = (5m - 3)(m + 1)$

9. $6r^2 + 11r - 2 = (6r - 1)(r + 2)$

11. $6y^2 - 7y - 5 = (3y - 5)(2y + 1)$

13. $6t^2 + 19t + 10 = (3t + 2)(2t + 5)$

15. $6 + 11x + 4x^2 = 4x^2 + 11x + 6 = (4x + 3)(x + 2)$

17. $4y^2 - 12y + 9 = (2y - 3)^2$

19. $8k^2 - 10k - 3 = (4k + 1)(2k - 3)$

21. $48z + 20z^2 - 5 = 20z^2 + 48z - 5 = (10z - 1)(2z + 5)$

23. $4m^2 + m - 14 = (4m - 7)(m + 2)$

25. $20p^2 - 9p - 20 = (5p + 4)(4p - 5)$

27. $3z^2 + 11z + 7$ is prime.

29. $36r^2 - 5r - 24 = (9r - 8)(4r + 3)$

31. $2x^2 + 5xy + 3y^2 = (2x + 3y)(x + y)$

33. $4p^2 - 8pq + 3q^2 = (2p - 3q)(2p - q)$

35. $10a^2 - 23ab + 9b^2 = (5a - 9b)(2a - b)$

37. $3r^2 + 14rs - 5s^2 = (3r - s)(r + 5s)$

39. $8m^2 - 2mn - 21n^2 = (4m - 7n)(2m + 3n)$

41. $12c^2 + 7cd - 12d^2 = (4c - 3d)(3c + 4d)$

43. $6x^2 + 3xy - 8y^2$ is prime.

45. $25p^2 - 30pq + 9q^2 = (5p - 3q)^2$

47. $12m^2 + 10n^2 - 23mn = 12m^2 - 23mn + 10n^2 = (4m - 5n)(3m - 2n)$

49. $10x^2 + 16x + 6 = 2(5x^2 + 8x + 3) = 2(5x + 3)(x + 1)$

51. $24y^2 - 28y + 8 = 4(6y^2 - 7y + 2) = 4(3y - 2)(2y - 1)$

53. $-4x^2 - 13x - 3 = -(4x^2 + 13x + 3) = -(4x + 1)(x + 3)$

55. $-12r^2 + 26r + 10 = -2(6r^2 - 13r - 5) = -2(3r + 1)(2r - 5)$

57. $9z^5 + 6z^4 + z^3 = z^3(9z^2 + 6z + 1) = z^3(3z + 1)^2$

59. $10p^2 - 210p + 100p^3 = 100p^3 + 10p^2 - 210p = 10p(10p^2 + p - 21) = 10p(5p - 7)(2p + 3)$

61. $24m^4n - 7m^3n^2 - 6m^2n^3 = m^2n(24m^2 - 7mn - 6n^2) = m^2n(8m + 3n)(3m - 2n)$

63. $12p^6q - 15p^5q + 30p^4q = 3p^4q(4p^2 - 5p + 10)$

65. $-18a^3b - 3a^2b^2 + 105ab^3 = -3ab(6a^2 + ab - 35b^2) = -3ab(3a - 7b)(2a + 5b)$

67. $2x^4 - x^2y^2 - 3y^4 = (2x^2 - 3y^2)(x^2 + y^2)$

69. $5x^6 + 8x^3y + 3y^2 = (5x^3 + 3y)(x^3 + y)$

71. $(3x - 4)(3x + 4) = (3x)^2 - 4^2 = 9x^2 - 16$

Problem Set 4.5, pp. 199-200

1. $x^2 - 4 = x^2 - 2^2 = (x + 2)(x - 2)$

3. $m^2 - n^2 = (m + n)(m - n)$

5. $9r^2 - 25 = (3r)^2 - 5^2 = (3r + 5)(3r - 5)$

7. $x^2 + 9$ is prime.

9. $y^2 - \dfrac{16}{81} = y^2 - (\dfrac{4}{9})^2 = (y + \dfrac{4}{9})(y - \dfrac{4}{9})$

11. $36 - z^2 = 6^2 - z^2 = (6 + z)(6 - z)$

13. $4p^2 - q^2 = (2p)^2 - q^2 = (2p + q)(2p - q)$

15. $m^2 + n^2$ is prime.

17. $121x^2 - 81y^2 = (11x)^2 - (9y)^2 = (11x + 9y)(11x - 9y)$

19. $r^2s^2 - 169 = (rs)^2 - 13^2 = (rs + 13)(rs - 13)$

21. $m^4 - 9 = (m^2)^2 - 3^2 = (m^2 + 3)(m^2 - 3)$

23. $7x^4 - 7 = 7(x^4 - 1) = 7(x^2 + 1)(x^2 - 1) = 7(x^2 + 1)(x + 1)(x - 1)$

25. $-3t^6 + 48t^2 = -3t^2(t^4 - 16) = -3t^2(t^2 + 4)(t^2 - 4)$

$$= -3t^2(t^2 + 4)(t + 2)(t - 2)$$

27. $324x^5 - 64xy^4 = 4x(81x^4 - 16y^4) = 4x(9x^2 + 4y^2)(9x^2 - 4y^2)$

$$= 4x(9x^2 + 4y^2)(3x + 2y)(3x - 2y)$$

29. $x^2 + 10x + 25 = x^2 + 2(x)(5) + 5^2 = (x + 5)^2$

31. $y^2 - 22y + 121 = y^2 - 2(y)(11) + 11^2 = (y - 11)^2$

33. $1 + 6r + 9r^2 = 9r^2 + 6r + 1 = (3r)^2 + 2(3r)(1) + 1^2 = (3r + 1)^2$

35. $4m^2 - 12m + 9 = (2m)^2 - 2(2m)(3) + 3^2 = (2m - 3)^2$

37. $x^2 + 10x + 16 = (x + 2)(x + 8)$

39. $y^2 - 7y + 49$ is prime.

41. $25a^2 + 20ab + 4b^2 = (5a)^2 + 2(5a)(2b) + (2b)^2 = (5a + 2b)^2$

43. $2p^2 - 40pq + 200q^2 = 2(p^2 - 20pq + 100q^2) = 2(p - 10q)^2$

45. $5x^3 + 30x^2y + 45xy^2 = 5x(x^2 + 6xy + 9y^2) = 5x(x + 3y)^2$

47. $x^3 + 27 = x^3 + 3^3 = (x + 3)(x^2 - x \cdot 3 + 3^2) = (x + 3)(x^2 - 3x + 9)$

49. $m^3 - 8 = m^3 - 2^3 = (m - 2)(m^2 + m \cdot 2 + 2^2) = (m - 2)(m^2 + 2m + 4)$

51. $125r^3 + 1 = (5r)^3 + 1^3 = (5r + 1)[(5r)^2 - (5r)1 + 1^2]$

$$= (5r + 1)(25r^2 - 5r + 1)$$

53. $1000r^3 + s^3 = (10r)^3 + s^3 = (10r + s)[(10r)^2 - (10r)s + s^2]$

$$= (10r + s)(100r^2 - 10rs + s^2)$$

55. $8y^3 - 125 = (2y)^3 - 5^3 = (2y - 5)[(2y)^2 + (2y)(5) + 5^2]$

$$= (2y - 5)(4y^2 + 10y + 25)$$

57. $27a^3 + 64b^3 = (3a)^3 + (4b)^3 = (3a + 4b)[(3a)^2 - (3a)(4b) + (4b)^2]$

$$= (3a + 4b)(9a^2 - 12ab + 16b^2)$$

59. $x^6 + 343y^3 = (x^2)^3 + (7y)^3 = (x^2 + 7y)[(x^2)^2 - (x^2)(7y) + (7y)^2]$

$$= (x^2 + 7y)(x^4 - 7x^2y + 49y^2)$$

61. $-3k^4 - 648k = -3k(k^3 + 216) = -3k(k^3 + 6^3)$

$$= -3k(k + 6)(k^2 - k \cdot 6 + 6^2)$$

$$= -3k(k + 6)(k^2 - 6k + 36)$$

63. $320p^6 - 625q^3 = 5(64p^6 - 125q^3) = 5[(4p^2)^3 - (5q)^3]$

$$= 5(4p^2 - 5q)[(4p^2)^2 + (4p^2)(5q) + (5q)^2]$$

$$= 5(4p^2 - 5q)(16p^4 + 20p^2q + 25q^2)$$

65. $8x^9 - 125y^3 = (2x^3)^3 - (5y)^3 = (2x^3 - 5y)[(2x^3)^2 + (2x^3)(5y) + (5y)^2]$

$$= (2x^3 - 5y)(4x^6 + 10x^3y + 25y^2)$$

67. $x^2 - (y + 3)^2 = [x + (y + 3)][x - (y + 3)] = (x + y + 3)(x - y - 3)$

69. $100x^4 - 140x^2y + 49y^2 = (10x^2 - 7y)^2$

71. $(x^2 + 4x + 4) - y^2 = (x + 2)^2 - y^2 = [(x + 2) + y][(x + 2) - y]$

$$= (x + 2 + y)(x + 2 - y)$$

73. $(x + 1)^3 - y^3 = [(x + 1) - y][(x + 1)^2 + (x + 1)y + y^2]$

$$= (x + 1 - y)[(x + 1)^2 + y(x + 1) + y^2]$$

Problem Set 4.6, p. 202

1. $x^2 - 5x - 14 = (x - 7)(x + 2)$

3. $y^2 - 64 = y^2 - 8^2 = (y + 8)(y - 8)$

5. $6z^2 + 18z - 3 = 3(2z^2 + 6z - 1)$

7. $4p^2 + 4 = 4(p^2 + 1)$

9. $xy - 7y + 2x - 14 = y(x - 7) + 2(x - 7) = (x - 7)(y + 2)$

11. $k^3 + 9k^2 + k + 9 = k^2(k + 9) + 1(k + 9) = (k + 9)(k^2 + 1)$

13. $36 - 12t + t^2 = t^2 - 12t + 36 = (t - 6)^2$

15. $18m - 45 = 9(2m - 5)$

17. $c^3 + 8d^3 = c^3 + (2d)^3 = (c + 2d)[c^2 - c \cdot 2d + (2d)^2]$

$$= (c + 2d)(c^2 - 2cd + 4d^2)$$

19. $-15r^3 + 10r^2 - 5r = -5r(3r^2 - 2r + 1)$

21. $a^3 - a^2 = a^2(a - 1)$

23. $16p^2 - 40pq + 25q^2 = (4p - 5q)^2$

25. $4x^2 + 9$ is prime.

27. $z^3 - 1 = z^3 - 1^3 = (z - 1)(z^2 + z \cdot 1 + 1^2) = (z - 1)(z^2 + z + 1)$

29. $2y^2 + 3y - 20 = (2y - 5)(y + 4)$

31. $12c^2 + cd - 6d^2 = (4c + 3d)(3c - 2d)$

33. $1 - 49k^2 = 1^2 - (7k)^2 = (1 + 7k)(1 - 7k)$

35. $2xy - 5y - 2x + 5 = y(2x - 5) - 1(2x - 5) = (2x - 5)(y - 1)$

37. $x^{11}y^3z^5 + x^4y^3z^7 - x^8y^3z^9 = x^4y^3z^5(x^7 + z^2 - x^4z^4)$

39. $4m^2 + 6mn - 70n^2 = 2(2m^2 + 3mn - 35n^2) = 2(2m - 7n)(m + 5n)$

41. $-12 - 11b + b^2 = b^2 - 11b - 12 = (b - 12)(b + 1)$

43. $r^3s^3 - 27 = (rs)^3 - 3^3 = (rs - 3)(r^2s^2 + 3rs + 9)$

45. $96a^2 - 150b^2 = 6(16a^2 - 25b^2) = 6(4a + 5b)(4a - 5b)$

47. $-4t^3 - 19t^2 + 30t = -t(4t^2 + 19t - 30) = -t(4t - 5)(t + 6)$

49. $4\pi R^2 - 4\pi r^2 = 4\pi(R^2 - r^2) = 4\pi(R + r)(R - r)$

51. $8p^4 + 125pq^3 = p(8p^3 + 125q^3) = p[(2p)^3 + (5q)^3]$

$$= p(2p + 5q)(4p^2 - 10pq + 25q^2)$$

53. $6x^7 - 486x^3 = 6x^3(x^4 - 81) = 6x^3(x^2 + 9)(x^2 - 9)$

$$= 6x^3(x^2 + 9)(x + 3)(x - 3)$$

55. $-12 - 29m - 14m^2 = -14m^2 - 29m - 12 = -(14m^2 + 29m + 12)$

$$= -(7m + 4)(2m + 3)$$

57. $x^3 + 3x^2 - 4x - 12 = x^2(x + 3) - 4(x + 3) = (x + 3)(x^2 - 4)$

$$= (x + 3)(x + 2)(x - 2)$$

59. $12a^2b + 24ab - 20a - 40 = 4(3a^2b + 6ab - 5a - 10)$

$$= 4[3ab(a + 2) - 5(a + 2)]$$

$$= 4(a + 2)(3ab - 5)$$

61. $121x^4 - 154x^2y^2 + 49y^4 = (11x^2 - 7y^2)^2$

63. $6x^6 + 7x^3y - 20y^2 = (3x^3 - 4y)(2x^3 + 5y)$

65. $m^3 - (n + 3)^3 = [m - (n + 3)][m^2 + m(n + 3) + (n + 3)^2]$

$$= (m - n - 3)[m^2 + m(n + 3) + (n + 3)^2]$$

67. $4p^6 - 4 = 4(p^6 - 1) = 4(p^3 + 1)(p^3 - 1)$

$$= 4(p + 1)(p^2 - p + 1)(p - 1)(p^2 + p + 1)$$

69. $x^2 - y^2 - 2y - 1 = x^2 - (y^2 + 2y + 1) = x^2 - (y + 1)^2$

$$= [x + (y + 1)][x - (y + 1)]$$

$$= (x + y + 1)(x - y - 1)$$

71. $a^2 - b^2 + 5a + 5b = (a + b)(a - b) + 5(a + b)$

$$= (a + b)(a - b + 5)$$

Problem Set 4.7, pp. 207-208

1. $(y - 1)(y + 3) = 0$

    $y - 1 = 0$ or $y + 3 = 0$

        $y = 1$ $y = -3$

3. $(m - 9)(m - 9) = 0$

    $m - 9 = 0$ or $m - 9 = 0$

        $m = 9$ $m = 9$

5. $3t(7t - 4) = 0$

    $3t = 0$ or $7t - 4 = 0$

      $t = 0$ $7t = 4$

                    $t = \dfrac{4}{7}$

7. $x^2 - 3x + 2 = 0$

    $(x - 1)(x - 2) = 0$

    $x - 1 = 0$ or $x - 2 = 0$

        $x = 1$ $x = 2$

9. $z^2 - 2z - 35 = 0$

    $(z - 7)(z + 5) = 0$

    $z - 7 = 0$ or $z + 5 = 0$

        $z = 7$ $z = -5$

11. $3m^2 + 2m - 1 = 0$

    $(3m - 1)(m + 1) = 0$

    $3m - 1 = 0$ or $m + 1 = 0$

        $3m = 1$ $m = -1$

         $m = \dfrac{1}{3}$

13. $2y^2 + 11y = 6$

    $2y^2 + 11y - 6 = 0$

    $(2y - 1)(y + 6) = 0$

    $2y - 1 = 0$ or $y + 6 = 0$

        $2y = 1$ $y = -6$

         $y = \dfrac{1}{2}$

15. $5r^2 - 18r = 8$

    $5r^2 - 18r - 8 = 0$

    $(5r + 2)(r - 4) = 0$

    $5r + 2 = 0$ or $r - 4 = 0$

        $5r = -2$ $r = 4$

         $r = -\dfrac{2}{5}$

17. $-t^2 + 6t - 9 = 0$

$-1(t^2 - 6t + 9) = -1(0)$

$t^2 - 6t + 9 = 0$

$(t - 3)(t - 3) = 0$

$t - 3 = 0$ or $t - 3 = 0$

$t = 3$ $t = 3$

19. $r^2 - 81 = 0$

$(r + 9)(r - 9) = 0$

$r + 9 = 0$ or $r - 9 = 0$

$r = -9$ $r = 9$

21. $p^2 + 8p = 0$

$p(p + 8) = 0$

$p = 0$ or $p + 8 = 0$

$p = -8$

23. $4m^2 = 25$

$4m^2 - 25 = 0$

$(2m + 5)(2m - 5) = 0$

$2m + 5 = 0$ or $2m - 5 = 0$

$2m = -5$ $2m = 5$

$m = -\dfrac{5}{2}$ $m = \dfrac{5}{2}$

25. $9t^2 + 4 = 12t$

$9t^2 - 12t + 4 = 0$

$(3t - 2)(3t - 2) = 0$

$3t - 2 = 0$ or $3t - 2 = 0$

$3t = 2$ $3t = 2$

$t = \dfrac{2}{3}$ $t = \dfrac{2}{3}$

27. $6x^2 = 0$

$6 \cdot x \cdot x = 0$

$x = 0$ or $x = 0$

29. $10x^2 + 46x + 24 = 0$

$2(5x^2 + 23x + 12) = 0$

$2(5x + 3)(x + 4) = 0$

$5x + 3 = 0$ or $x + 4 = 0$

$5x = -3$ $x = -4$

$x = -\dfrac{3}{5}$

31. $8y^2 - 4y = 0$

$4y(2y - 1) = 0$

$4y = 0$ or $2y - 1 = 0$

$y = 0$ $2y = 1$

$y = \dfrac{1}{2}$

33.
$$1 + 12m = -36m^2$$
$$36m^2 + 12m + 1 = 0$$
$$(6m + 1)(6m + 1) = 0$$
$$6m + 1 = 0 \text{ or } 6m + 1 = 0$$
$$6m = -1 \qquad 6m = -1$$
$$m = -\frac{1}{6} \qquad m = -\frac{1}{6}$$

35.
$$5r^2 = 20r$$
$$5r^2 - 20r = 0$$
$$5r(r - 4) = 0$$
$$5r = 0 \text{ or } r - 4 = 0$$
$$r = 0 \qquad r = 4$$

37.
$$m(m + 1) = 2$$
$$m^2 + m = 2$$
$$m^2 + m - 2 = 0$$
$$(m + 2)(m - 1) = 0$$
$$m + 2 = 0 \text{ or } m - 1 = 0$$
$$m = -2 \qquad m = 1$$

39.
$$(p - 3)(p + 1) = 12$$
$$p^2 - 2p - 3 = 12$$
$$p^2 - 2p - 15 = 0$$
$$(p - 5)(p + 3) = 0$$
$$p - 5 = 0 \text{ or } p + 3 = 0$$
$$p = 5 \qquad p = -3$$

41.
$$(2y + 1)^2 = (y + 2)^2 + 9$$
$$4y^2 + 4y + 1 = y^2 + 4y + 4 + 9$$
$$4y^2 + 4y + 1 = y^2 + 4y + 13$$
$$3y^2 - 12 = 0$$
$$3(y^2 - 4) = 0$$
$$3(y + 2)(y - 2) = 0$$
$$y + 2 = 0 \text{ or } y - 2 = 0$$
$$y = -2 \qquad y = 2$$

43.
$$2k(k - 6) = (3k + 4)(k - 5)$$
$$2k^2 - 12k = 3k^2 - 11k - 20$$
$$0 = k^2 + k - 20$$
$$0 = (k + 5)(k - 4)$$
$$k + 5 = 0 \text{ or } k - 4 = 0$$
$$k = -5 \qquad k = 4$$

45. $$(2r - 1)^2 - (r + 3)^2 = 0$$
$$4r^2 - 4r + 1 - (r^2 + 6r + 9) = 0$$
$$4r^2 - 4r + 1 - r^2 - 6r - 9 = 0$$
$$3r^2 - 10r - 8 = 0$$
$$(3r + 2)(r - 4) = 0$$
$$3r + 2 = 0 \quad \text{or} \quad r - 4 = 0$$
$$3r = -2 \qquad r = 4$$
$$r = -\frac{2}{3}$$

47. $$(x - 2)(x + 7)(4x - 3) = 0$$
$$x - 2 = 0 \quad \text{or} \quad x + 7 = 0 \quad \text{or} \quad 4x - 3 = 0$$
$$x = 2 \qquad x = -7 \qquad 4x = 3$$
$$x = \frac{3}{4}$$

49. $$(y - 4)(y^2 + 7y - 8) = 0$$
$$(y - 4)(y + 8)(y - 1) = 0$$
$$y - 4 = 0 \quad \text{or} \quad y + 8 = 0 \quad \text{or} \quad y - 1 = 0$$
$$y = 4 \qquad y = -8 \qquad y = 1$$

51. $$p^3 - 49p = 0$$
$$p(p^2 - 49) = 0$$
$$p(p + 7)(p - 7) = 0$$
$$p = 0 \quad \text{or} \quad p + 7 = 0 \quad \text{or} \quad p - 7 = 0$$
$$p = -7 \qquad p = 7$$

53. $$m^3 - 13m^2 = 0$$
$$m^2(m - 13) = 0$$
$$m^2 = 0 \quad \text{or} \quad m - 13 = 0$$
$$m = 0 \qquad m = 13$$

55. $$x^3 + 5x^2 = 6x$$
$$x^3 + 5x^2 - 6x = 0$$
$$x(x^2 + 5x - 6) = 0$$
$$x(x + 6)(x - 1) = 0$$
$$x = 0 \quad \text{or} \quad x + 6 = 0 \quad \text{or} \quad x - 1 = 0$$
$$x = -6 \qquad x = 1$$

57.
$$r^4 = r^3 + 20r^2$$
$$r^4 - r^3 - 20r^2 = 0$$
$$r^2(r^2 - r - 20) = 0$$
$$r^2(r - 5)(r + 4) = 0$$
$$r^2 = 0 \text{ or } r - 5 = 0 \text{ or } r + 4 = 0$$
$$r = 0 \qquad r = 5 \qquad r = -4$$

59. x = the number
$$4x + x^2 = 21$$
$$x^2 + 4x - 21 = 0$$
$$(x + 7)(x - 3) = 0$$
$$x + 7 = 0 \text{ or } x - 3 = 0$$
$$x = -7 \qquad x = 3$$

61. x = the number
$$x^2 = 4 + 3x$$
$$x^2 - 3x - 4 = 0$$
$$(x - 4)(x + 1) = 0$$
$$x - 4 = 0 \text{ or } x + 1 = 0$$
$$x = 4 \qquad x = -1$$

Problem Set 4.8, pp. 211-213

1.   x = first integer
        x + 2 = second integer
$$x(x + 2) = 63$$
$$x^2 + 2x = 63$$
$$x^2 + 2x - 63 = 0$$
$$(x + 9)(x - 7) = 0$$
$$x + 9 = 0 \text{ or } x - 7 = 0$$
$$x = -9 \qquad x = 7$$
$$x + 2 = -7 \qquad x + 2 = 9$$

3.   x = first integer
        x + 1 = second integer
$$x^2 + (x + 1)^2 = 41$$
$$x^2 + x^2 + 2x + 1 = 41$$
$$2x^2 + 2x - 40 = 0$$
$$2(x^2 + x - 20) = 0$$
$$2(x + 5)(x - 4) = 0$$
$$x + 5 = 0 \text{ or } x - 4 = 0$$
$$x = -5 \qquad x = 4$$
$$x + 1 = 5$$

5. x = first number

   $2x$ = second number

$$x^2 - 2x = 5(x + 2x)$$

$$x^2 - 2x = 5(3x)$$

$$x^2 - 2x = 15x$$

$$x^2 - 17x = 0$$

$$x(x - 17) = 0$$

$$\cancel{x = 0} \quad \text{or} \quad x - 17 = 0$$

$$x = 17$$

$$2x = 34$$

7. x = first integer

   $x + 1$ = second integer

$$(x + x + 1)^2 = 221 + 2x(x + 1)$$

$$(2x + 1)^2 = 221 + 2x^2 + 2x$$

$$4x^2 + 4x + 1 = 221 + 2x^2 + 2x$$

$$2x^2 + 2x - 220 = 0$$

$$2(x^2 + x - 110) = 0$$

$$2(x + 11)(x - 10) = 0$$

$$x + 11 = 0 \quad \text{or} \quad x - 10 = 0$$

$$x = -11 \qquad \cancel{x = 10}$$

$$x + 1 = -10$$

9. x = width

   $3x + 1$ = length

$$\text{Area} = 30$$

$$x(3x + 1) = 30$$

$$3x^2 + x = 30$$

$$3x^2 + x - 30 = 0$$

$$(x - 3)(3x + 10) = 0$$

$$x - 3 = 0 \quad \text{or} \quad 3x + 10 = 0$$

$$x = 3\,\text{m} \qquad 3x = -10$$

$$3x + 1 = 10 \text{ m} \qquad \cancel{x = -\frac{10}{3}}$$

11. x = width

   $6 - 2x$ = length

$$\text{Area} = 4$$

$$x(6 - 2x) = 4$$

$$6x - 2x^2 = 4$$

$$0 = 2x^2 - 6x + 4$$

$$0 = 2(x^2 - 3x + 2)$$

$$0 = 2(x - 1)(x - 2)$$

$$x - 1 = 0 \quad \text{or} \quad x - 2 = 0$$

$$x = 1 \text{ mi} \qquad x = 2 \text{ mi}$$

$$6 - 2x = 4 \text{ mi} \qquad 6 - 2x = 2 \text{ mi}$$

13.

	Width	Length	Area
Original rectangle	x	$x + 3$	$x(x + 3)$
New rectangle	$x + 4$	$x + 7$	$(x + 4)(x + 7)$

New area = original area + 44

$$(x + 4)(x + 7) = x(x + 3) + 44$$

$$x^2 + 11x + 28 = x^2 + 3x + 44$$

$$8x = 16$$

$$x = 2 \text{ cm}$$

$$x + 3 = 5 \text{ cm}$$

15. x = height ladder reaches

$$a^2 + b^2 = c^2$$

$$x^2 + 30^2 = 50^2$$

$$x^2 + 900 = 2500$$

$$x^2 - 1600 = 0$$

$$(x + 40)(x - 40) = 0$$

$$x + 40 = 0 \quad \text{or} \quad x - 40 = 0$$

$$x = -40 \qquad\qquad x = 40 \text{ ft}$$

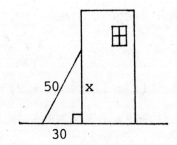

17.   x = length of cable

$$c^2 = a^2 + b^2$$

$$x^2 = 7^2 + 24^2$$

$$x^2 = 49 + 576$$

$$x^2 = 625$$

$$x^2 - 625 = 0$$

$$(x + 25)(x - 25) = 0$$

$$x + 25 = 0 \quad \text{or} \quad x - 25 = 0$$

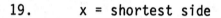 -25   x = 25 ft

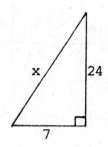

19.   x = shortest side

x + 1 = middle side

x + 2 = longest side

$$a^2 + b^2 = c^2$$

$$x^2 + (x + 1)^2 = (x + 2)^2$$

$$x^2 + x^2 + 2x + 1 = x^2 + 4x + 4$$

$$x^2 - 2x - 3 = 0$$

$$(x - 3)(x + 1) = 0$$

$$x - 3 = 0 \quad \text{or} \quad x + 1 = 0$$

x = 3 in.   -1

x + 1 = 4 in.

x + 2 = 5 in.

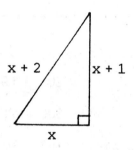

21. x = width

  $2x - 1$ = length

  $2x + 1$ = diagonal

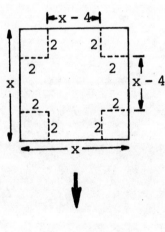

$$a^2 + b^2 = c^2$$

$$x^2 + (2x - 1)^2 = (2x + 1)^2$$

$$x^2 + 4x^2 - 4x + 1 = 4x^2 + 4x + 1$$

$$x^2 - 8x = 0$$

$$x(x - 8) = 0$$

~~$x = 0$~~ or $x - 8 = 0$

$$x = 8 \text{ yd}$$

$$2x - 1 = 15 \text{ yd}$$

23. x = length of each side

  Volume = 128

  $\ell wh = 128$

  $(x - 4)(x - 4)2 = 128$

  $(x - 4)^2 = 64$

  $x^2 - 8x + 16 = 64$

  $x^2 - 8x - 48 = 0$

  $(x - 12)(x + 4) = 0$

  $x - 12 = 0$ or $x + 4 = 0$

  $x = 12 \text{ in.}$ ~~$x = -4$~~

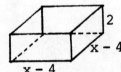

25. $d = 16t^2$

       $144 = 16t^2$

       $9 = t^2$

       $0 = t^2 - 9$

       $0 = (t + 3)(t - 3)$

       $t + 3 = 0$ or $t - 3 = 0$

              ~~$t = -3$~~ $t = 3$ sec

CHAPTER 5

RATIONAL EXPRESSIONS

Problem Set 5.1, pp. 222-223

1. $2x = 0$

    $x = 0$

3. $x - 6 = 0$

    $x = 6$

5. $r^2 - 25 = 0$

    $(r + 5)(r - 5) = 0$

    $r + 5 = 0$ or $r - 5 = 0$

    $r = -5$ $r = 5$

7. $y^2 - 4y - 21 = 0$

    $(y - 7)(y + 3) = 0$

    $y - 7 = 0$ or $y + 3 = 0$

    $y = 7$ $y = -3$

9. $3 = 0$

    No solution

11. $t^2 + 4 = 0$

    No solution

13. $\dfrac{36x^3}{9x} = \dfrac{2 \cdot 2 \cdot 3 \cdot 3 \cdot x \cdot x \cdot x}{3 \cdot 3 \cdot x} = \dfrac{2 \cdot 2 \cdot x \cdot x}{1} = 4x^2$

15. $\dfrac{-6m}{21m^2} = \dfrac{-1 \cdot 2 \cdot 3 \cdot m}{3 \cdot 7 \cdot m \cdot m} = \dfrac{-2}{7m}$

17. $\dfrac{-22p^3}{-99p^3} = \dfrac{-1 \cdot 2 \cdot 11 \cdot p \cdot p \cdot p}{-1 \cdot 3 \cdot 3 \cdot 11 \cdot p \cdot p \cdot p} = \dfrac{2}{9}$

19. $\dfrac{30a^2b^4}{-24a^3b} = \dfrac{2 \cdot 3 \cdot 5 \cdot a \cdot a \cdot b \cdot b \cdot b \cdot b}{-1 \cdot 2 \cdot 2 \cdot 2 \cdot 3 \cdot a \cdot a \cdot a \cdot b} = \dfrac{5 \cdot b \cdot b \cdot b}{-1 \cdot 2 \cdot 2 \cdot a} = -\dfrac{5b^3}{4a}$

21. $\dfrac{12x + 18}{6} = \dfrac{6(2x + 3)}{6} = \dfrac{2x + 3}{1} = 2x + 3$

23. $\dfrac{2p - 6}{5p - 15} = \dfrac{2(p - 3)}{5(p - 3)} = \dfrac{2}{5}$

25. $\dfrac{3m + 3}{9m + 9} = \dfrac{3(m + 1)}{9(m + 1)} = \dfrac{3(m + 1)}{3 \cdot 3(m + 1)} = \dfrac{1}{3}$

27. $\dfrac{10y^2}{25y^2 - 5y} = \dfrac{2 \cdot 5 \cdot y \cdot y}{5y(5y - 1)} = \dfrac{2y}{5y - 1}$

29. $\dfrac{x^2 + 8x}{8x + 64} = \dfrac{x(x + 8)}{8(x + 8)} = \dfrac{x}{8}$

31. $\dfrac{t - 4}{t^2 - 16} = \dfrac{t - 4}{(t + 4)(t - 4)} = \dfrac{1}{t + 4}$

33. $\dfrac{x^2 - 25}{x^2 + 2x - 35} = \dfrac{(x + 5)(x - 5)}{(x + 7)(x - 5)} = \dfrac{x + 5}{x + 7}$

35. $\dfrac{(r + s)^2}{(r + s)^6} = \dfrac{(r + s)^2}{(r + s)^2(r + s)^4} = \dfrac{1}{(r + s)^4}$

37. $\dfrac{(c + d)^2}{c^2 - d^2} = \dfrac{(c + d)(c + d)}{(c + d)(c - d)} = \dfrac{c + d}{c - d}$

39. $\dfrac{z^2 - 4z + 3}{z^2 - 2z + 1} = \dfrac{(z - 1)(z - 3)}{(z - 1)(z - 1)} = \dfrac{z - 3}{z - 1}$

41. $\dfrac{2m^2 - 9mn - 5n^2}{m^2 - 25n^2} = \dfrac{(2m + n)(m - 5n)}{(m + 5n)(m - 5n)} = \dfrac{2m + n}{m + 5n}$

43. $\dfrac{8x^2 + 2x - 6}{4x^2 + 5x - 6} = \dfrac{2(4x^2 + x - 3)}{(4x - 3)(x + 2)} = \dfrac{2(4x - 3)(x + 1)}{(4x - 3)(x + 2)} = \dfrac{2(x + 1)}{x + 2}$

45. $\dfrac{x + 3}{3 + x} = \dfrac{x + 3}{x + 3} = 1$

47. $\dfrac{x - 3}{3 - x} = \dfrac{1(x - 3)}{-1(x - 3)} = \dfrac{1}{-1} = -1$

49. $\dfrac{2r - 5}{5 - 2r} = \dfrac{1(2r - 5)}{-1(2r - 5)} = \dfrac{1}{-1} = -1$

51. $\dfrac{a^2 - ab - 2b^2}{6b^2 - ab - a^2} = \dfrac{(a - 2b)(a + b)}{-(a^2 + ab - 6b^2)} = \dfrac{(a - 2b)(a + b)}{-(a - 2b)(a + 3b)} = -\dfrac{a + b}{a + 3b}$

53. $\dfrac{4 + 2(t + 3)}{5t - 7(t + 2)} = \dfrac{4 + 2t + 6}{5t - 7t - 14} = \dfrac{2t + 10}{-2t - 14} = \dfrac{2(t + 5)}{-2(t + 7)} = -\dfrac{t + 5}{t + 7}$

55. $\dfrac{2(x + 3) - (x + 9)}{x^2 - 9} = \dfrac{2x + 6 - x - 9}{(x + 3)(x - 3)} = \dfrac{x - 3}{(x + 3)(x - 3)} = \dfrac{1}{x + 3}$

57. $\dfrac{(p - 5)p + 6}{(p - 2)p - 3} = \dfrac{p^2 - 5p + 6}{p^2 - 2p - 3} = \dfrac{(p - 2)(p - 3)}{(p + 1)(p - 3)} = \dfrac{p - 2}{p + 1}$

59. $\dfrac{(m + 2)3 - 6}{(m + 2)m} = \dfrac{3m + 6 - 6}{(m + 2)m} = \dfrac{3m}{(m + 2)m} = \dfrac{3}{m + 2}$

61. $\dfrac{x^3 + 27}{x + 3} = \dfrac{(x + 3)(x^2 - 3x + 9)}{x + 3} = \dfrac{x^2 - 3x + 9}{1} = x^2 - 3x + 9$

63. $\dfrac{m^3 + m^2 + 4m + 4}{m^2 - 2m - 3} = \dfrac{m^2(m + 1) + 4(m + 1)}{(m + 1)(m - 3)} = \dfrac{(m + 1)(m^2 + 4)}{(m + 1)(m - 3)} = \dfrac{m^2 + 4}{m - 3}$

65. $a = \dfrac{180n - 360}{n} = \dfrac{180(6) - 360}{6} = \dfrac{1080 - 360}{6} = \dfrac{720}{6} = 120°$

67. $r = \dfrac{100x}{15 + x} = \dfrac{100(22.5)}{15 + 22.5} = \dfrac{2250}{37.5} = 60$

Problem Set 5.2, pp. 227-228

1. $\dfrac{m^2}{14} \cdot \dfrac{10}{m} = \dfrac{m \cdot m}{2 \cdot 7} \cdot \dfrac{2 \cdot 5}{m} = \dfrac{5m}{7}$

3. $\dfrac{a^3}{9b} \cdot \dfrac{27}{ab} = \dfrac{a \cdot a \cdot a}{3 \cdot 3 \cdot b} \cdot \dfrac{3 \cdot 3 \cdot 3}{ab} = \dfrac{3a^2}{b^2}$

5. $\dfrac{y}{3} \div \dfrac{3}{y} = \dfrac{y}{3} \cdot \dfrac{y}{3} = \dfrac{y^2}{9}$

7. $\dfrac{-25k^{10}}{9k^7} \div \dfrac{20k^8}{6k^4} = \dfrac{-25k^{10}}{9k^7} \cdot \dfrac{6k^4}{20k^8} = -\dfrac{5}{6k}$

9. $\dfrac{36d}{3c} \div (4cd) = \dfrac{36d}{3c} \cdot \dfrac{1}{4cd} = \dfrac{3}{c^2}$

11. $\dfrac{(2r)^2}{15r^4} \cdot \dfrac{5r^3}{2(3r)^2} \cdot \dfrac{(3r)^3}{20r} = \dfrac{4r^2}{15r^4} \cdot \dfrac{5r^3}{18r^2} \cdot \dfrac{27r^3}{20r} = \dfrac{r}{10}$

13. $\dfrac{(3x)^2 y^5}{3(z^2 y)^4} \div \dfrac{6(xy)^3}{(-3z^2 y)^3} = \dfrac{9x^2 y^5}{3z^8 y^4} \div \dfrac{6x^3 y^3}{-27z^6 y^3} = \dfrac{9x^2 y^5}{3z^8 y^4} \cdot \dfrac{-27z^6 y^3}{6x^3 y^3}$

$$= -\dfrac{27y}{2xz^2}$$

15. $\dfrac{x^2}{6} \div \dfrac{x^3}{4} \div \dfrac{x}{y} = \left(\dfrac{x^2}{6} \div \dfrac{x^3}{4}\right) \div \dfrac{x}{y} = \left(\dfrac{x^2}{6} \cdot \dfrac{4}{x^3}\right) \div \dfrac{x}{y} = \dfrac{2}{3x} \div \dfrac{x}{y}$

$$= \dfrac{2}{3x} \cdot \dfrac{y}{x}$$

$$= \dfrac{2y}{3x^2}$$

17. $\dfrac{7a + 14}{5} \cdot \dfrac{15a}{6a + 12} = \dfrac{7(a + 2)}{5} \cdot \dfrac{15a}{6(a + 2)} = \dfrac{7a}{2}$

19. $\dfrac{x^2 - 9}{5x} \cdot \dfrac{10}{2x - 6} = \dfrac{(x + 3)(x - 3)}{5x} \cdot \dfrac{10}{2(x - 3)} = \dfrac{x + 3}{x}$

21. $\dfrac{m - 5}{m} \cdot \dfrac{m^3}{5 - m} = \dfrac{m - 5}{m} \cdot \dfrac{m^3}{-1(m - 5)} = \dfrac{m^2}{-1} = -m^2$

23. $\dfrac{4t + 6}{t^2 + t} \div \dfrac{6t + 9}{t} = \dfrac{4t + 6}{t^2 + t} \cdot \dfrac{t}{6t + 9} = \dfrac{2(2t + 3)}{t(t + 1)} \cdot \dfrac{t}{3(2t + 3)} = \dfrac{2}{3(t + 1)}$

25. $\dfrac{x^4 + x^3}{y + xy} \div \dfrac{x^3}{y^2} = \dfrac{x^4 + x^3}{y + xy} \cdot \dfrac{y^2}{x^3} = \dfrac{x^3(x + 1)}{y(1 + x)} \cdot \dfrac{y^2}{x^3} = \dfrac{y}{1} = y$

27. $\dfrac{p^2 - 6p}{p^2 + 6p + 5} \cdot \dfrac{p^2 + 7p + 10}{p^2 - 4p - 12} = \dfrac{p(p - 6)}{(p + 1)(p + 5)} \cdot \dfrac{(p + 2)(p + 5)}{(p + 2)(p - 6)} = \dfrac{p}{p + 1}$

29. $\dfrac{m^2 + 3m - 10}{m^2 - 25} \cdot \dfrac{m^2 - 2m - 15}{8m^2 - 16m} = \dfrac{(m + 5)(m - 2)}{(m + 5)(m - 5)} \cdot \dfrac{(m - 5)(m + 3)}{8m(m - 2)} = \dfrac{m + 3}{8m}$

31. $\dfrac{(c + d)^2}{8c} \cdot \dfrac{24c}{c^2 - d^2} = \dfrac{(c + d)(c + d)}{8c} \cdot \dfrac{24c}{(c + d)(c - d)} = \dfrac{3(c + d)}{c - d}$

33. $\dfrac{x}{y - 2} \div \dfrac{y - 6}{x} = \dfrac{x}{y - 2} \cdot \dfrac{x}{y - 6} = \dfrac{x^2}{(y - 2)(y - 6)}$

35. $\dfrac{t + 2}{t - 1} \div \dfrac{1}{t^2 - 1} = \dfrac{t + 2}{t - 1} \cdot \dfrac{t^2 - 1}{1} = \dfrac{t + 2}{t - 1} \cdot \dfrac{(t + 1)(t - 1)}{1} = (t + 2)(t + 1)$

37. $\dfrac{9z^2 - 1}{6z + 2} \div \dfrac{3z - 1}{2} = \dfrac{(3z + 1)(3z - 1)}{2(3z + 1)} \cdot \dfrac{2}{3z - 1} = 1$

39. $\dfrac{6y^2 + 7y - 20}{2y^2 + 9y + 10} \div \dfrac{3y^2 - y - 4}{y^2 - y - 2} = \dfrac{6y^2 + 7y - 20}{2y^2 + 9y + 10} \cdot \dfrac{y^2 - y - 2}{3y^2 - y - 4}$

$$= \dfrac{(3y - 4)(2y + 5)}{(2y + 5)(y + 2)} \cdot \dfrac{(y - 2)(y + 1)}{(3y - 4)(y + 1)}$$

$$= \dfrac{y - 2}{y + 2}$$

41. $\dfrac{7a + 35b}{a^2 + ab} \div \dfrac{a^2 + 6ab + 5b^2}{a^2 - b^2} = \dfrac{7a + 35b}{a^2 + ab} \cdot \dfrac{a^2 - b^2}{a^2 + 6ab + 5b^2}$

$$= \dfrac{7(a + 5b)}{a(a + b)} \cdot \dfrac{(a + b)(a - b)}{(a + 5b)(a + b)}$$

$$= \dfrac{7(a - b)}{a(a + b)}$$

43. $\dfrac{c^2 + c - 6}{c^2 - 6c + 8} \div \dfrac{c^2 + 2c - 8}{c^2 - c - 12} = \dfrac{c^2 + c - 6}{c^2 - 6c + 8} \cdot \dfrac{c^2 - c - 12}{c^2 + 2c - 8}$

$$= \dfrac{(c + 3)(c - 2)}{(c - 4)(c - 2)} \cdot \dfrac{(c - 4)(c + 3)}{(c + 4)(c - 2)}$$

$$= \dfrac{(c + 3)^2}{(c - 2)(c + 4)}$$

45. $\dfrac{36 - r^2}{2r + 12} \div (r^2 - 6r) = \dfrac{36 - r^2}{2r + 12} \cdot \dfrac{1}{r^2 - 6r} = \dfrac{(6 + r)(6 - r)}{2(r + 6)} \cdot \dfrac{1}{-r(6 - r)}$

$$= -\dfrac{1}{2r}$$

47. $\dfrac{12}{x} \div \dfrac{x + 3}{x - 3} \cdot \dfrac{3x^2 + 9x}{x^2 - 9} = \left(\dfrac{12}{x} \div \dfrac{x + 3}{x - 3}\right) \cdot \dfrac{3x^2 + 9x}{x^2 - 9}$

$$= \dfrac{12}{x} \cdot \dfrac{x - 3}{x + 3} \cdot \dfrac{3x(x + 3)}{(x + 3)(x - 3)}$$

$$= \dfrac{36}{x + 3}$$

49. $\dfrac{x^2 + 10x + 25}{2x^2 + 10x} \cdot \dfrac{20x}{2x^2 + 3x - 2} \div \dfrac{2x + 10}{x + 2}$

$$= \left(\dfrac{x^2 + 10x + 25}{2x^2 + 10x} \cdot \dfrac{20x}{2x^2 + 3x - 2}\right) \cdot \dfrac{x + 2}{2x + 10}$$

$$= \dfrac{(x + 5)(x + 5)}{2x(x + 5)} \cdot \dfrac{20x}{(2x - 1)(x + 2)} \cdot \dfrac{x + 2}{2(x + 5)}$$

$$= \dfrac{5}{2x - 1}$$

51. $$\frac{x^2 + 8x + 16}{5x^2 + 19x - 4} \div \frac{9x + 36}{3x^2 - 6x} \cdot \frac{15x^2 + 2x - 1}{x^2 - 4}$$

$$= \left(\frac{x^2 + 8x + 16}{5x^2 + 19x - 4} \div \frac{9x + 36}{3x^2 - 6x}\right) \cdot \frac{15x^2 + 2x - 1}{x^2 - 4}$$

$$= \frac{(x + 4)(x + 4)}{(5x - 1)(x + 4)} \cdot \frac{3x(x - 2)}{9(x + 4)} \cdot \frac{(5x - 1)(3x + 1)}{(x + 2)(x - 2)}$$

$$= \frac{x(3x + 1)}{3(x + 2)}$$

53. $$\frac{pq + 4p + 2q + 8}{pq + 4p + 3q + 12} \cdot \frac{p^2 - 9}{12p^2 + 24p} = \frac{p(q + 4) + 2(q + 4)}{p(q + 4) + 3(q + 4)} \cdot \frac{(p + 3)(p - 3)}{12p(p + 2)}$$

$$= \frac{(q + 4)(p + 2)}{(q + 4)(p + 3)} \cdot \frac{(p + 3)(p - 3)}{12p(p + 2)}$$

$$= \frac{p - 3}{12p}$$

Problem Set 5.3, pp. 231-232

1. $$\frac{a}{3} + \frac{a}{3} = \frac{a + a}{3} = \frac{2a}{3}$$

3. $$\frac{5x}{8} + \frac{x}{8} = \frac{5x + x}{8} = \frac{6x}{8} = \frac{3x}{4}$$

5. $$\frac{1}{p} + \frac{1}{p} = \frac{1 + 1}{p} = \frac{2}{p}$$

7. $$\frac{1}{6y} + \frac{5}{6y} = \frac{1 + 5}{6y} = \frac{6}{6y} = \frac{1}{y}$$

9. $$\frac{9}{10r} - \frac{7}{10r} = \frac{9 - 7}{10r} = \frac{2}{10r} = \frac{1}{5r}$$

11. $$\frac{2p}{p + 5} + \frac{10}{p + 5} = \frac{2p + 10}{p + 5} = \frac{2(p + 5)}{p + 5} = \frac{2}{1} = 2$$

13. $\dfrac{b + 3}{b} - \dfrac{3}{b} = \dfrac{b + 3 - 3}{b} = \dfrac{b}{b} = 1$

15. $\dfrac{m + 3}{7x} + \dfrac{m - 2}{7x} = \dfrac{m + 3 + m - 2}{7x} = \dfrac{2m + 1}{7x}$

17. $\dfrac{r + 8}{r^2} - \dfrac{1}{r^2} = \dfrac{r + 8 - 1}{r^2} = \dfrac{r + 7}{r^2}$

19. $\dfrac{x}{x^2} + \dfrac{2}{x^2} = \dfrac{x + 2}{x^2}$

21. $\dfrac{t^2}{t - 3} - \dfrac{9}{t - 3} = \dfrac{t^2 - 9}{t - 3} = \dfrac{(t + 3)(t - 3)}{t - 3} = t + 3$

23. $\dfrac{m + 5}{m - 1} - \dfrac{m + 2}{m - 1} = \dfrac{m + 5 - (m + 2)}{m - 1} = \dfrac{m + 5 - m - 2}{m - 1} = \dfrac{3}{m - 1}$

25. $\dfrac{3p}{p^2 + 5p - 14} - \dfrac{6}{p^2 + 5p - 14} = \dfrac{3p - 6}{p^2 + 5p - 14} = \dfrac{3(p - 2)}{(p + 7)(p - 2)} = \dfrac{3}{p + 7}$

27. $\dfrac{a^2 + b^2}{a + b} + \dfrac{2ab}{a + b} = \dfrac{a^2 + b^2 + 2ab}{a + b} = \dfrac{a^2 + 2ab + b^2}{a + b} = \dfrac{(a + b)(a + b)}{a + b}$

$$= a + b$$

29. $\dfrac{3y + 5}{y^2 + 2y} + \dfrac{y + 3}{y^2 + 2y} = \dfrac{3y + 5 + y + 3}{y^2 + 2y} = \dfrac{4y + 8}{y^2 + 2y} = \dfrac{4(y + 2)}{y(y + 2)} = \dfrac{4}{y}$

31. $\dfrac{6x + 7}{x - 4} - \dfrac{5x + 3}{x - 4} = \dfrac{6x + 7 - (5x + 3)}{x - 4} = \dfrac{6x + 7 - 5x - 3}{x - 4} = \dfrac{x + 4}{x - 4}$

33. $\dfrac{x(x + 5)}{(x - 5)(x + 5)} - \dfrac{5(x - 5)}{(x + 5)(x - 5)} = \dfrac{x(x + 5) - 5(x - 5)}{(x - 5)(x + 5)} = \dfrac{x^2 + 5x - 5x + 25}{(x - 5)(x + 5)}$

$$= \dfrac{x^2 + 25}{(x - 5)(x + 5)}$$

35. $\dfrac{2(x + 8)}{x^2 - 2x - 15} - \dfrac{5(x + 1)}{x^2 - 2x - 15} = \dfrac{2(x + 8) - 5(x + 1)}{x^2 - 2x - 15} = \dfrac{2x + 16 - 5x - 5}{x^2 - 2x - 15}$

$$= \dfrac{-3x + 11}{x^2 - 2x - 15}$$

$$= \dfrac{-3x + 11}{(x - 5)(x + 3)}$$

37. $\dfrac{(m + 2)(m + 3)}{(m + 4)(m + 3)} + \dfrac{(m + 2)(m - 1)}{(m + 3)(m + 4)} = \dfrac{(m + 2)(m + 3) + (m + 2)(m - 1)}{(m + 4)(m + 3)}$

$$= \dfrac{m^2 + 5m + 6 + m^2 + m - 2}{(m + 4)(m + 3)}$$

$$= \dfrac{2m^2 + 6m + 4}{(m + 4)(m + 3)}$$

$$= \dfrac{2(m^2 + 3m + 2)}{(m + 4)(m + 3)}$$

$$= \dfrac{2(m + 1)(m + 2)}{(m + 4)(m + 3)}$$

39. $\dfrac{y^2 + y + 7}{(y + 5)(y + 2)} - \dfrac{(y - 4)(y + 2)}{(y + 5)(y + 2)} = \dfrac{y^2 + y + 7 - (y - 4)(y + 2)}{(y + 5)(y + 2)}$

$$= \dfrac{y^2 + y + 7 - (y^2 - 2y - 8)}{(y + 5)(y + 2)}$$

$$= \dfrac{y^2 + y + 7 - y^2 + 2y + 8}{(y + 5)(y + 2)}$$

$$= \dfrac{3y + 15}{(y + 5)(y + 2)}$$

$$= \dfrac{3(y + 5)}{(y + 5)(y + 2)}$$

$$= \dfrac{3}{y + 2}$$

41. $\dfrac{8}{m-2} + \dfrac{3}{2-m} = \dfrac{8}{m-2} + \dfrac{3}{2-m} \cdot \dfrac{-1}{-1} = \dfrac{8}{m-2} + \dfrac{-3}{m-2}$

$$= \dfrac{8 + (-3)}{m-2}$$

$$= \dfrac{5}{m-2}$$

43. $\dfrac{4}{x-y} - \dfrac{5}{y-x} = \dfrac{4}{x-y} - \dfrac{5}{y-x} \cdot \dfrac{-1}{-1} = \dfrac{4}{x-y} - \dfrac{-5}{x-y}$

$$= \dfrac{4 - (-5)}{x-y}$$

$$= \dfrac{9}{x-y}$$

45. $\dfrac{2r+1}{3-r} - \dfrac{5r-8}{r-3} = \dfrac{-1}{-1} \cdot \dfrac{2r+1}{3-r} - \dfrac{5r-8}{r-3} = \dfrac{-2r-1}{r-3} - \dfrac{5r-8}{r-3}$

$$= \dfrac{-2r - 1 - 5r + 8}{r-3}$$

$$= \dfrac{-7r + 7}{r-3}$$

$$= \dfrac{-7(r-1)}{r-3}$$

47. $\dfrac{x^2+x-1}{(x-1)^2} - \dfrac{x(x+1)}{(x-1)^2} + \dfrac{x}{(x-1)^2} = \dfrac{x^2 + x - 1 - x(x+1) + x}{(x-1)^2}$

$$= \dfrac{x^2 + x - 1 - x^2 - x + x}{(x-1)^2}$$

$$= \dfrac{x-1}{(x-1)^2}$$

$$= \dfrac{1}{x-1}$$

Problem Set 5.4, pp. 235-236

1. $y^2 = y^2$

    $y^6 = y^6$

    LCD $= y^6$

3. $9 = 3^2$

    $z = z$

    LCD $= 3^2 \cdot z = 9z$

5. $1 = 1$

    $3k = 3 \cdot k$

    LCD $= 3 \cdot k = 3k$

7. $5p = 5 \cdot p$

    $9p = 3^2 \cdot p$

    LCD $= 3^2 \cdot 5 \cdot p = 45p$

9. $6m^2 = 2 \cdot 3 \cdot m^2$

    $18m = 2 \cdot 3^2 \cdot m$

    LCD $= 2 \cdot 3^2 \cdot m^2 = 18m^2$

11. $a^3b = a^3 \cdot b$

    $a^2b^2 = a^2 \cdot b^2$

    LCD $= a^3 \cdot b^2 = a^3b^2$

13. $12xy^4 = 2^2 \cdot 3 \cdot x \cdot y^4$

    $45x^3y^2 = 3^2 \cdot 5 \cdot x^3 \cdot y^2$

    LCD $= 2^2 \cdot 3^2 \cdot 5 \cdot x^3 \cdot y^4$

    $= 180x^3y^4$

15. $m + 4 = m + 4$

    $3m + 12 = 3(m + 4)$

    LCD $= 3(m + 4)$

17. $2r - 6 = 2(r - 3)$

    $3r - 9 = 3(r - 3)$

    LCD $= 2 \cdot 3 \cdot (r - 3)$

    $= 6(r - 3)$

19. $k^2 + k = k(k + 1)$

    $k + 1 = k + 1$

    LCD $= k(k + 1)$

21. $x = x$

    $x - 1 = x - 1$

    LCD $= x(x - 1)$

23. $t^2 - 4t = t(t - 4)$

    $t^2 + 5t = t(t + 5)$

    LCD $= t(t - 4)(t + 5)$

25. $18z + 12 = 6(3z + 2)$

$27z + 18 = 9(3z + 2)$

$LCD = 18(3z + 2)$

27. $y + 3 = y + 3$

$y + 5 = y + 5$

$LCD = (y + 3)(y + 5)$

29. $5m - 3 = 5m - 3$

$10m + 9 = 10m + 9$

$LCD = (5m - 3)(10m + 9)$

31. $4r + 8 = 4(r + 2)$

$3r - 6 = 3(r - 2)$

$LCD = 12(r + 2)(r - 2)$

33. $a + 7 = a + 7$

$1 = 1$

$LCD = a + 7$

35. $x^2 - 4 = (x + 2)(x - 2)$

$x + 2 = x + 2$

$LCD = (x + 2)(x - 2)$

37. $3x + 18 = 3(x + 6)$

$x^2 - 36 = (x + 6)(x - 6)$

$LCD = 3(x + 6)(x - 6)$

39. $x^2 - y^2 = (x + y)(x - y)$

$x^2 + 2xy + y^2 = (x + y)^2$

$LCD = (x + y)^2(x - y)$

41. $p^2 + 10p = p(p + 10)$

$10p = 10p$

$LCD = 10p(p + 10)$

43. $2x + 8 = 2(x + 4)$

$x^2 + 3x - 4 = (x + 4)(x - 1)$

$LCD = 2(x + 4)(x - 1)$

45. $r^2 - 25 = (r + 5)(r - 5)$

$r^2 - 10r + 25 = (r - 5)^2$

$LCD = (r + 5)(r - 5)^2$

47. $m^2 - 8m - 9 = (m - 9)(m + 1)$

$m^2 - 5m - 6 = (m - 6)(m + 1)$

$LCD = (m - 9)(m + 1)(m - 6)$

49. $6a^2 - 11a - 10 = (3a + 2)(2a - 5)$

$3a^2 - 4a - 4 = (3a + 2)(a - 2)$

$LCD = (3a + 2)(2a - 5)(a - 2)$

$165 = 3 \cdot 5 \cdot 11$

$12 = 2^2 \cdot 3$

$30 = 2 \cdot 3 \cdot 5$

660 years

51. x is the smallest number that is divisible by both 24 and 36; that is,

x is the LCD of 24 and 36.

$24 = 2^3 \cdot 3$

$36 = 2^2 \cdot 3^2$

LCD $= 2^3 \cdot 3^2 = 72$ oranges

Problem Set 5.5, pp. 241-243

1. $\dfrac{x}{10} + \dfrac{x}{5} = \dfrac{x}{10} + \dfrac{x}{5} \cdot \dfrac{2}{2} = \dfrac{x}{10} + \dfrac{2x}{10} = \dfrac{3x}{10}$

3. $\dfrac{a}{2} - \dfrac{2}{a} = \dfrac{a}{2} \cdot \dfrac{a}{a} - \dfrac{2}{a} \cdot \dfrac{2}{2} = \dfrac{a^2}{2a} - \dfrac{4}{2a} = \dfrac{a^2 - 4}{2a}$

5. $\dfrac{1}{2x} + \dfrac{1}{9x} = \dfrac{1}{2x} \cdot \dfrac{9}{9} + \dfrac{1}{9x} \cdot \dfrac{2}{2} = \dfrac{9}{18x} + \dfrac{2}{18x} = \dfrac{11}{18x}$

7. $\dfrac{5}{12m} - \dfrac{1}{8m} = \dfrac{5}{12m} \cdot \dfrac{2}{2} - \dfrac{1}{8m} \cdot \dfrac{3}{3} = \dfrac{10}{24m} - \dfrac{3}{24m} = \dfrac{7}{24m}$

9. $\dfrac{1}{r^2} + \dfrac{3}{r} = \dfrac{1}{r^2} + \dfrac{3}{r} \cdot \dfrac{r}{r} = \dfrac{1}{r^2} + \dfrac{3r}{r^2} = \dfrac{1 + 3r}{r^2}$

11. $\dfrac{4}{y} - \dfrac{2}{y^2} = \dfrac{4}{y} \cdot \dfrac{y}{y} - \dfrac{2}{y^2} = \dfrac{4y}{y^2} - \dfrac{2}{y^2} = \dfrac{4y - 2}{y^2}$

13. $\dfrac{8}{a^2} + \dfrac{6}{ab} = \dfrac{8}{a^2} \cdot \dfrac{b}{b} + \dfrac{6}{ab} \cdot \dfrac{a}{a} = \dfrac{8b}{a^2b} + \dfrac{6a}{a^2b} = \dfrac{8b + 6a}{a^2b}$

15. $\dfrac{3}{12x^2y^4} - \dfrac{5}{18x^2y^5} = \dfrac{3}{12x^2y^4} \cdot \dfrac{3y}{3y} - \dfrac{5}{18x^2y^5} \cdot \dfrac{2}{2} = \dfrac{9y}{36x^2y^5} - \dfrac{10}{36x^2y^5} = \dfrac{9y - 10}{36x^2y^5}$

17. $\dfrac{2}{xy^2} - \dfrac{3}{x^2y} + \dfrac{1}{xy} = \dfrac{2}{xy^2} \cdot \dfrac{x}{x} - \dfrac{3}{x^2y} \cdot \dfrac{y}{y} + \dfrac{1}{xy} \cdot \dfrac{xy}{xy} = \dfrac{2x}{x^2y^2} - \dfrac{3y}{x^2y^2} + \dfrac{xy}{x^2y^2}$

$$= \dfrac{2x - 3y + xy}{x^2y^2}$$

19. $\quad 7 + \dfrac{1}{x} = \dfrac{7}{1} \cdot \dfrac{x}{x} + \dfrac{1}{x} = \dfrac{7x}{x} + \dfrac{1}{x} = \dfrac{7x + 1}{x}$

21. $\quad \dfrac{1}{4t} + t = \dfrac{1}{4t} + \dfrac{t}{1} \cdot \dfrac{4t}{4t} = \dfrac{1}{4t} + \dfrac{4t^2}{4t} = \dfrac{1 + 4t^2}{4t}$

23. $\quad a - \dfrac{a^2}{a + 4} = \dfrac{a}{1} \cdot \dfrac{a + 4}{a + 4} - \dfrac{a^2}{a + 4} = \dfrac{a^2 + 4a}{a + 4} - \dfrac{a^2}{a + 4} = \dfrac{a^2 + 4a - a^2}{a + 4}$

$$= \dfrac{4a}{a + 4}$$

25. $\quad \dfrac{1}{x} - \dfrac{1}{x - 4} = \dfrac{1}{x} \cdot \dfrac{x - 4}{x - 4} - \dfrac{1}{x - 4} \cdot \dfrac{x}{x} = \dfrac{x - 4}{x(x - 4)} - \dfrac{x}{x(x - 4)} = \dfrac{x - 4 - x}{x(x - 4)}$

$$= \dfrac{-4}{x(x - 4)}$$

27. $\quad \dfrac{m}{m + 6} + \dfrac{1}{m} = \dfrac{m}{m + 6} \cdot \dfrac{m}{m} + \dfrac{1}{m} \cdot \dfrac{m + 6}{m + 6} = \dfrac{m^2}{m(m + 6)} + \dfrac{m + 6}{m(m + 6)} = \dfrac{m^2 + m + 6}{m(m + 6)}$

29. $\quad \dfrac{3}{a + 2} + \dfrac{8}{a - 4} = \dfrac{3}{a + 2} \cdot \dfrac{a - 4}{a - 4} + \dfrac{8}{a - 4} \cdot \dfrac{a + 2}{a + 2} = \dfrac{3a - 12}{(a + 2)(a - 4)} + \dfrac{8a + 16}{(a - 4)(a + 2)}$

$$= \dfrac{3a - 12 + 8a + 16}{(a + 2)(a - 4)}$$

$$= \dfrac{11a + 4}{(a + 2)(a - 4)}$$

31. $\quad \dfrac{x}{x - 3} - \dfrac{1}{x + 3} = \dfrac{x}{x - 3} \cdot \dfrac{x + 3}{x + 3} - \dfrac{1}{x + 3} \cdot \dfrac{x - 3}{x - 3} = \dfrac{x^2 + 3x}{(x - 3)(x + 3)} - \dfrac{x - 3}{(x + 3)(x - 3)}$

$$= \dfrac{x^2 + 3x - (x - 3)}{(x - 3)(x + 3)}$$

$$= \dfrac{x^2 + 3x - x + 3}{(x - 3)(x + 3)}$$

$$= \dfrac{x^2 + 2x + 3}{(x - 3)(x + 3)}$$

33. $\dfrac{t-4}{t+8} - \dfrac{t+1}{t-2} = \dfrac{t-4}{t+8} \cdot \dfrac{t-2}{t-2} - \dfrac{t+1}{t-2} \cdot \dfrac{t+8}{t+8} = \dfrac{t^2 - 6t + 8}{(t+8)(t-2)} - \dfrac{t^2 + 9t + 8}{(t-2)(t+8)}$

$$= \dfrac{t^2 - 6t + 8 - (t^2 + 9t + 8)}{(t+8)(t-2)}$$

$$= \dfrac{t^2 - 6t + 8 - t^2 - 9t - 8}{(t+8)(t-2)}$$

$$= \dfrac{-15t}{(t+8)(t-2)}$$

35. $\dfrac{10}{3m+12} + \dfrac{1}{m+4} = \dfrac{10}{3(m+4)} + \dfrac{1}{m+4} \cdot \dfrac{3}{3} = \dfrac{10}{3(m+4)} + \dfrac{3}{3(m+4)} = \dfrac{13}{3(m+4)}$

37. $\dfrac{6}{r^2 - 1} + \dfrac{3}{r+1} = \dfrac{6}{(r+1)(r-1)} + \dfrac{3}{r+1} \cdot \dfrac{r-1}{r-1} = \dfrac{6}{(r+1)(r-1)} + \dfrac{3r-3}{(r+1)(r-1)}$

$$= \dfrac{6 + 3r - 3}{(r+1)(r-1)}$$

$$= \dfrac{3r + 3}{(r+1)(r-1)}$$

$$= \dfrac{3(r+1)}{(r+1)(r-1)}$$

$$= \dfrac{3}{r-1}$$

39. $\dfrac{22}{p^2 - 4} + \dfrac{11}{2p + 4} = \dfrac{22}{(p + 2)(p - 2)} \cdot \dfrac{2}{2} + \dfrac{11}{2(p + 2)} \cdot \dfrac{p - 2}{p - 2}$

$$= \dfrac{44}{2(p + 2)(p - 2)} + \dfrac{11p - 22}{2(p + 2)(p - 2)}$$

$$= \dfrac{44 + 11p - 22}{2(p + 2)(p - 2)}$$

$$= \dfrac{11p + 22}{2(p + 2)(p - 2)}$$

$$= \dfrac{11(p + 2)}{2(p + 2)(p - 2)}$$

$$= \dfrac{11}{2(p - 2)}$$

41. $\dfrac{2}{t^2 - 3t} + \dfrac{3}{5t} = \dfrac{2}{t(t - 3)} \cdot \dfrac{5}{5} + \dfrac{3}{5t} \cdot \dfrac{t - 3}{t - 3} = \dfrac{10}{5t(t - 3)} + \dfrac{3t - 9}{5t(t - 3)}$

$$= \dfrac{3t + 1}{5t(t - 3)}$$

43. $\dfrac{6k}{(k - 1)^2} - \dfrac{2}{k^2 - 1} = \dfrac{6k}{(k - 1)^2} \cdot \dfrac{k + 1}{k + 1} - \dfrac{2}{(k + 1)(k - 1)} \cdot \dfrac{k - 1}{k - 1}$

$$= \dfrac{6k^2 + 6k}{(k - 1)^2(k + 1)} - \dfrac{2k - 2}{(k + 1)(k - 1)^2}$$

$$= \dfrac{6k^2 + 6k - (2k - 2)}{(k - 1)^2(k + 1)}$$

$$= \dfrac{6k^2 + 6k - 2k + 2}{(k - 1)^2(k + 1)}$$

$$= \dfrac{6k^2 + 4k + 2}{(k - 1)^2(k + 1)}$$

45. $\dfrac{r}{r^2 + 4r + 4} - \dfrac{r+3}{r^2 + 3r + 2} = \dfrac{r}{(r+2)^2} \cdot \dfrac{r+1}{r+1} - \dfrac{r+3}{(r+1)(r+2)} \cdot \dfrac{r+2}{r+2}$

$$= \dfrac{r^2 + r}{(r+2)^2(r+1)} - \dfrac{r^2 + 5r + 6}{(r+1)(r+2)^2}$$

$$= \dfrac{r^2 + r - (r^2 + 5r + 6)}{(r+2)^2(r+1)}$$

$$= \dfrac{r^2 + r - r^2 - 5r - 6}{(r+2)^2(r+1)}$$

$$= \dfrac{-4r - 6}{(r+2)^2(r+1)}$$

47. $\dfrac{m-3}{m^2 + 4m - 5} + \dfrac{m-2}{m^2 + 11m + 30} = \dfrac{m-3}{(m+5)(m-1)} \cdot \dfrac{m+6}{m+6} + \dfrac{m-2}{(m+5)(m+6)} \cdot \dfrac{m-1}{m-1}$

$$= \dfrac{m^2 + 3m - 18}{(m+5)(m-1)(m+6)} + \dfrac{m^2 - 3m + 2}{(m+5)(m+6)(m-1)}$$

$$= \dfrac{m^2 + 3m - 18 + m^2 - 3m + 2}{(m+5)(m-1)(m+6)}$$

$$= \dfrac{2m^2 - 16}{(m+5)(m-1)(m+6)}$$

49. $\dfrac{9x - 20}{x^2 + x - 12} - \dfrac{6x - 13}{x^2 - x - 6} = \dfrac{9x - 20}{(x+4)(x-3)} \cdot \dfrac{x+2}{x+2} - \dfrac{6x - 13}{(x-3)(x+2)} \cdot \dfrac{x+4}{x+4}$

$$= \dfrac{9x^2 - 2x - 40 - (6x^2 + 11x - 52)}{(x+4)(x-3)(x+2)}$$

$$= \dfrac{3x^2 - 13x + 12}{(x+4)(x-3)(x+2)}$$

$$= \dfrac{(3x - 4)(x - 3)}{(x+4)(x-3)(x+2)}$$

$$= \dfrac{3x - 4}{(x+4)(x+2)}$$

51. $\dfrac{1}{x - 3} + \dfrac{2}{x + 3} - \dfrac{6}{x^2 - 9} = \dfrac{1}{x - 3} \cdot \dfrac{x + 3}{x + 3} + \dfrac{2}{x + 3} \cdot \dfrac{x - 3}{x - 3} - \dfrac{6}{(x + 3)(x - 3)}$

$$= \dfrac{x + 3 + 2x - 6 - 6}{(x - 3)(x + 3)}$$

$$= \dfrac{3x - 9}{(x - 3)(x + 3)}$$

$$= \dfrac{3(x - 3)}{(x - 3)(x + 3)}$$

$$= \dfrac{3}{x + 3}$$

53. $\dfrac{2y}{y - 6} + \dfrac{1}{y + 4} + \dfrac{10}{y^2 - 2y - 24}$

$$= \dfrac{2y}{y - 6} \cdot \dfrac{y + 4}{y + 4} + \dfrac{1}{y + 4} \cdot \dfrac{y - 6}{y - 6} + \dfrac{10}{(y - 6)(y + 4)}$$

$$= \dfrac{2y^2 + 8y + y - 6 + 10}{(y - 6)(y + 4)}$$

$$= \dfrac{2y^2 + 9y + 4}{(y - 6)(y + 4)}$$

$$= \dfrac{(2y + 1)(y + 4)}{(y - 6)(y + 4)}$$

$$= \dfrac{2y + 1}{y - 6}$$

55. $\dfrac{4z}{z^2 - 16} - \dfrac{2}{z} - \dfrac{2}{z - 4}$

$$= \frac{4z}{(z + 4)(z - 4)} \cdot \frac{z}{z} - \frac{2}{z} \cdot \frac{(z + 4)(z - 4)}{(z + 4)(z - 4)} - \frac{2}{z - 4} \cdot \frac{z(z + 4)}{z(z + 4)}$$

$$= \frac{4z^2 - 2(z^2 - 16) - 2z(z + 4)}{z(z + 4)(z - 4)}$$

$$= \frac{4z^2 - 2z^2 + 32 - 2z^2 - 8z}{z(z + 4)(z - 4)}$$

$$= \frac{-8z + 32}{z(z + 4)(z - 4)}$$

$$= \frac{-8(z - 4)}{z(z + 4)(z - 4)}$$

$$= \frac{-8}{z(z + 4)}$$

57. $\left[\dfrac{-x}{2 - x} - \dfrac{3}{(x - 2)^2}\right]\dfrac{x^3 - 8}{x^3 - 3x^2 + 4x - 12}$

$$= \left[\frac{x}{x - 2} - \frac{3}{(x - 2)^2}\right]\frac{x^3 - 2^3}{x^2(x - 3) + 4(x - 3)}$$

$$= \left[\frac{x}{x - 2} \cdot \frac{x - 2}{x - 2} - \frac{3}{(x - 2)^2}\right]\frac{(x - 2)(x^2 + 2x + 4)}{(x - 3)(x^2 + 4)}$$

$$= \left[\frac{x^2 - 2x - 3}{(x - 2)^2}\right]\frac{(x - 2)(x^2 + 2x + 4)}{(x - 3)(x^2 + 4)}$$

$$= \frac{(x + 1)(x - 3)}{(x - 2)^2} \cdot \frac{(x - 2)(x^2 + 2x + 4)}{(x - 3)(x^2 + 4)}$$

$$= \frac{(x + 1)(x^2 + 2x + 4)}{(x - 2)(x^2 + 4)}$$

59. $\dfrac{a}{b} + \dfrac{c}{d} = \dfrac{a}{b} \cdot \dfrac{d}{d} + \dfrac{c}{d} \cdot \dfrac{b}{b} = \dfrac{ad}{bd} + \dfrac{bc}{bd} = \dfrac{ad + bc}{bd}$

Problem Set 5.6, pp. 247-249

1. $\dfrac{1 + \dfrac{1}{5}}{2} = \dfrac{(1 + \dfrac{1}{5})5}{(2)5} = \dfrac{5 + 1}{10} = \dfrac{6}{10} = \dfrac{3}{5}$

3. $\dfrac{\dfrac{1}{3} + \dfrac{1}{6}}{\dfrac{3}{4} + \dfrac{1}{3}} = \dfrac{(\dfrac{1}{3} + \dfrac{1}{6})12}{(\dfrac{3}{4} + \dfrac{1}{3})12} = \dfrac{4 + 2}{9 + 4} = \dfrac{6}{13}$

5. $\dfrac{\dfrac{x + 1}{x}}{\dfrac{x - 1}{x}} = \dfrac{x + 1}{x} \div \dfrac{x - 1}{x} = \dfrac{x + 1}{x} \cdot \dfrac{x}{x - 1} = \dfrac{x + 1}{x - 1}$

7. $\dfrac{\dfrac{2r^3}{s^2}}{\dfrac{6r^3}{s^3}} = \dfrac{2r^3}{s^2} \div \dfrac{6r^3}{s^3} = \dfrac{2r^3}{s^2} \cdot \dfrac{s^3}{6r^3} = \dfrac{s}{3}$

9. $\dfrac{\dfrac{1}{m^2 - 16}}{\dfrac{1}{m + 4}} = \dfrac{1}{m^2 - 16} \div \dfrac{1}{m + 4} = \dfrac{1}{(m + 4)(m - 4)} \cdot \dfrac{m + 4}{1} = \dfrac{1}{m - 4}$

11. $\dfrac{\dfrac{1}{b} + a}{\dfrac{1}{a} + b} = \dfrac{(\dfrac{1}{b} + a)ab}{(\dfrac{1}{a} + b)ab} = \dfrac{a + a^2b}{b + ab^2} = \dfrac{a(1 + ab)}{b(1 + ab)} = \dfrac{a}{b}$

13. $\dfrac{\dfrac{1}{a} + \dfrac{1}{b}}{\dfrac{a + b}{b}} = \dfrac{\left(\dfrac{1}{a} + \dfrac{1}{b}\right)ab}{\left(\dfrac{a + b}{b}\right)ab} = \dfrac{b + a}{(a + b)a} = \dfrac{1}{a}$

15. $\dfrac{1 - \dfrac{x}{y}}{1 - \dfrac{x^2}{y^2}} = \dfrac{\left(1 - \dfrac{x}{y}\right)y^2}{\left(1 - \dfrac{x^2}{y^2}\right)y^2} = \dfrac{y^2 - xy}{y^2 - x^2} = \dfrac{y(y - x)}{(y + x)(y - x)} = \dfrac{y}{y + x}$

17. $\dfrac{\dfrac{1}{xy} + 1}{\dfrac{1}{x} - \dfrac{y}{x^2}} = \dfrac{\left(\dfrac{1}{xy} + 1\right)x^2 y}{\left(\dfrac{1}{x} - \dfrac{y}{x^2}\right)x^2 y} = \dfrac{x + x^2 y}{xy - y^2} = \dfrac{x(1 + xy)}{y(x - y)}$

19. $\dfrac{\dfrac{1}{ab} - \dfrac{1}{b}}{\dfrac{1}{b} - \dfrac{1}{ab}} = \dfrac{\left(\dfrac{1}{ab} - \dfrac{1}{b}\right)ab}{\left(\dfrac{1}{b} - \dfrac{1}{ab}\right)ab} = \dfrac{1 - a}{a - 1} = -1$

21. $\dfrac{1 + \dfrac{2}{x}}{\dfrac{1}{4} - \dfrac{1}{x^2}} = \dfrac{\left(1 + \dfrac{2}{x}\right)4x^2}{\left(\dfrac{1}{4} - \dfrac{1}{x^2}\right)4x^2} = \dfrac{4x^2 + 8x}{x^2 - 4} = \dfrac{4x(x + 2)}{(x + 2)(x - 2)} = \dfrac{4x}{x - 2}$

23. $\dfrac{t^2 - 1}{1 - \dfrac{1}{t}} = \dfrac{(t^2 - 1)t}{\left(1 - \dfrac{1}{t}\right)t} = \dfrac{(t + 1)(t - 1)t}{t - 1} = (t + 1)t$

25. $\dfrac{\dfrac{3}{4z} - \dfrac{2}{3z^2}}{\dfrac{1}{2z} + \dfrac{5}{6z^2}} = \dfrac{\left(\dfrac{3}{4z} - \dfrac{2}{3z^2}\right)12z^2}{\left(\dfrac{1}{2z} + \dfrac{5}{6z^2}\right)12z^2} = \dfrac{9z - 8}{6z + 10} = \dfrac{9z - 8}{2(3z + 5)}$

27. $\dfrac{1 - \dfrac{9}{p^2}}{1 - \dfrac{1}{p} - \dfrac{6}{p^2}} = \dfrac{(1 - \dfrac{9}{p^2})p^2}{(1 - \dfrac{1}{p} - \dfrac{6}{p^2})p^2} = \dfrac{p^2 - 9}{p^2 - p - 6} = \dfrac{(p + 3)(p - 3)}{(p + 2)(p - 3)} = \dfrac{p + 3}{p + 2}$

29. $\dfrac{2 + \dfrac{5}{x} - \dfrac{12}{x^2}}{3 + \dfrac{11}{x} - \dfrac{4}{x^2}} = \dfrac{(2 + \dfrac{5}{x} - \dfrac{12}{x^2})x^2}{(3 + \dfrac{11}{x} - \dfrac{4}{x^2})x^2} = \dfrac{2x^2 + 5x - 12}{3x^2 + 11x - 4} = \dfrac{(2x - 3)(x + 4)}{(3x - 1)(x + 4)}$

$$= \dfrac{2x - 3}{3x - 1}$$

31. $\dfrac{1}{1 + \dfrac{1}{x + 1}} = \dfrac{1(x + 1)}{(1 + \dfrac{1}{x + 1})(x + 1)} = \dfrac{x + 1}{x + 1 + 1} = \dfrac{x + 1}{x + 2}$

33. $\dfrac{\dfrac{5}{m + 2} + 1}{1 - \dfrac{5}{m + 2}} = \dfrac{(\dfrac{5}{m + 2} + 1)(m + 2)}{(1 - \dfrac{5}{m + 2})(m + 2)} = \dfrac{5 + m + 2}{m + 2 - 5} = \dfrac{m + 7}{m - 3}$

35. $\dfrac{k - 7 + \dfrac{5}{k - 1}}{k - 3 + \dfrac{1}{k - 1}} = \dfrac{(k - 7 + \dfrac{5}{k - 1})(k - 1)}{(k - 3 + \dfrac{1}{k - 1})(k - 1)} = \dfrac{(k - 7)(k - 1) + 5}{(k - 3)(k - 1) + 1}$

$$= \dfrac{k^2 - 8k + 7 + 5}{k^2 - 4k + 3 + 1}$$

$$= \dfrac{k^2 - 8k + 12}{k^2 - 4k + 4}$$

$$= \dfrac{(k - 2)(k - 6)}{(k - 2)(k - 2)}$$

$$= \dfrac{k - 6}{k - 2}$$

37. $\dfrac{\dfrac{1}{y} - \dfrac{1}{y+2}}{\dfrac{1}{y} + \dfrac{1}{y+2}} = \dfrac{\left(\dfrac{1}{y} - \dfrac{1}{y+2}\right)y(y+2)}{\left(\dfrac{1}{y} + \dfrac{1}{y+2}\right)y(y+2)} = \dfrac{y+2-y}{y+2+y} = \dfrac{2}{2y+2}$

$$= \dfrac{2}{2(y+1)}$$

$$= \dfrac{1}{y+1}$$

39. $1 + \dfrac{1}{1 - \dfrac{1}{1+1}} = 1 + \dfrac{1}{1 - \dfrac{1}{2}} = 1 + \dfrac{1}{\dfrac{1}{2}} = 1 + \dfrac{1(2)}{\dfrac{1}{2}(2)} = 1 + \dfrac{2}{1} = 3$

41. $\dfrac{a^{-1}}{b^{-2}} = \dfrac{\dfrac{1}{a}}{\dfrac{1}{b^2}} = \dfrac{1}{a} \div \dfrac{1}{b^2} = \dfrac{1}{a} \cdot \dfrac{b^2}{1} = \dfrac{b^2}{a}$

43. $\dfrac{x^{-2} - 1}{x^{-1} - 1} = \dfrac{\dfrac{1}{x^2} - 1}{\dfrac{1}{x} - 1} = \dfrac{\left(\dfrac{1}{x^2} - 1\right)x^2}{\left(\dfrac{1}{x} - 1\right)x^2} = \dfrac{1 - x^2}{x - x^2} = \dfrac{(1+x)(1-x)}{x(1-x)} = \dfrac{1+x}{x}$

45. $\dfrac{m-2}{8m^{-2} - m} = \dfrac{m-2}{\dfrac{8}{m^2} - m} = \dfrac{(m-2)m^2}{\left(\dfrac{8}{m^2} - m\right)m^2} = \dfrac{(m-2)m^2}{8 - m^3}$

$$= \dfrac{(m-2)m^2}{(2-m)(4 + 2m + m^2)}$$

$$= \dfrac{-m^2}{m^2 + 2m + 4}$$

47. $E = \dfrac{\dfrac{x}{2}}{x + \dfrac{1}{2}} = \dfrac{\dfrac{1}{2}x}{x + \dfrac{1}{2}} = \dfrac{\dfrac{1}{2}\left(\dfrac{3}{4}\right)}{\dfrac{3}{4} + \dfrac{1}{2}} = \dfrac{\dfrac{3}{8}}{\dfrac{5}{4}} = \dfrac{3}{8} \div \dfrac{5}{4}$

$$= \dfrac{3}{8} \cdot \dfrac{4}{5}$$

$$= \dfrac{3}{10}$$

49. $r = \dfrac{2}{\dfrac{1}{a} + \dfrac{1}{b}} = \dfrac{2}{\dfrac{1}{2} + \dfrac{1}{6}} = \dfrac{(2)6}{\left(\dfrac{1}{2} + \dfrac{1}{6}\right)6} = \dfrac{12}{3 + 1} = \dfrac{12}{4} = 3$ mph

51. $r = \dfrac{2}{\dfrac{1}{a} + \dfrac{1}{b}} = \dfrac{2}{\dfrac{1}{4} + \dfrac{1}{20}} = \dfrac{(2)20}{\left(\dfrac{1}{4} + \dfrac{1}{20}\right)20} = \dfrac{40}{5 + 1} = \dfrac{40}{6} = 6\dfrac{2}{3}$ mph

Problem Set 5.7, pp. 254-255

1. $\dfrac{x}{3} + \dfrac{x}{6} = 2$

$6 \cdot \dfrac{x}{3} + 6 \cdot \dfrac{x}{6} = 6 \cdot 2$

$2x + x = 12$

$3x = 12$

$x = 4$

3. $\dfrac{3r}{2} - \dfrac{3r}{4} = 6$

$4 \cdot \dfrac{3r}{2} - 4 \cdot \dfrac{3r}{4} = 4 \cdot 6$

$6r - 3r = 24$

$3r = 24$

$r = 8$

5. $\dfrac{y}{3} - 2 = \dfrac{y}{4}$

$12 \cdot \dfrac{y}{3} - 12 \cdot 2 = 12 \cdot \dfrac{y}{4}$

$4y - 24 = 3y$

$-24 = -y$

$24 = y$

7. $t + \dfrac{3}{2} = \dfrac{3t}{4} - \dfrac{t}{8}$

$8 \cdot t + 8 \cdot \dfrac{3}{2} = 8 \cdot \dfrac{3t}{4} - 8 \cdot \dfrac{t}{8}$

$8t + 12 = 6t - t$

$8t + 12 = 5t$

$12 = -3t$

$-4 = t$

9. $$\frac{1}{4x} - \frac{1}{2x} = \frac{1}{8}$$

$$8x \cdot \frac{1}{4x} - 8x \cdot \frac{1}{2x} = 8x \cdot \frac{1}{8}$$

$$2 - 4 = x$$

$$-2 = x$$

11. $$\frac{1}{2m} + \frac{1}{3m} = \frac{1}{6}$$

$$6m \cdot \frac{1}{2m} + 6m \cdot \frac{1}{3m} = 6m \cdot \frac{1}{6}$$

$$3 + 2 = m$$

$$5 = m$$

13. $$\frac{4}{y} + \frac{3}{5} = 1$$

$$5y \cdot \frac{4}{y} + 5y \cdot \frac{3}{5} = 5y \cdot 1$$

$$20 + 3y = 5y$$

$$20 = 2y$$

$$10 = y$$

15. $$7 - \frac{5}{2p} = \frac{3}{p} + \frac{3}{2}$$

$$2p \cdot 7 - 2p \cdot \frac{5}{2p} = 2p \cdot \frac{3}{p} + 2p \cdot \frac{3}{2}$$

$$14p - 5 = 6 + 3p$$

$$11p = 11$$

$$p = 1$$

17. $$1 + \frac{1}{x} - \frac{6}{x^2} = 0$$

$$x^2 \cdot 1 + x^2 \cdot \frac{1}{x} - x^2 \cdot \frac{6}{x^2} = x^2 \cdot 0$$

$$x^2 + x - 6 = 0$$

$$(x + 3)(x - 2) = 0$$

$$x + 3 = 0 \quad \text{or} \quad x - 2 = 0$$

$$x = -3 \qquad x = 2$$

19. $$\frac{x + 4}{x - 5} = \frac{2}{3}$$

$$3(x - 5)\frac{x + 4}{x - 5} = 3(x - 5)\frac{2}{3}$$

$$3(x + 4) = (x - 5)2$$

$$3x + 12 = 2x - 10$$

$$x = -22$$

21. $\dfrac{z + 4}{z - 1} = \dfrac{5}{z - 1}$

$(z - 1)\dfrac{z + 4}{z - 1} = (z - 1)\dfrac{5}{z - 1}$

$z + 4 = 5$

$z = 1$

Since z = 1 makes both denominators 0, there is no solution.

23. $\dfrac{r^2}{r + 3} = \dfrac{9}{r + 3}$

$(r + 3)\dfrac{r^2}{r + 3} = (r + 3)\dfrac{9}{r + 3}$

$r^2 = 9$

$r^2 - 9 = 0$

$(r + 3)(r - 3) = 0$

$r + 3 = 0$ or $r - 3 = 0$

$r = -3$ $r = 3$

Since r = -3 makes both denominators 0, the only solution is r = 3.

25. $\dfrac{x + 3}{2} + \dfrac{x + 1}{4} = 4$

$4 \cdot \dfrac{x + 3}{2} + 4 \cdot \dfrac{x + 1}{4} = 4 \cdot 4$

$2(x + 3) + x + 1 = 16$

$2x + 6 + x + 1 = 16$

$3x + 7 = 16$

$3x = 9$

$x = 3$

27. $\dfrac{2p - 1}{3} - 4 = \dfrac{p - 6}{2}$

$6 \cdot \dfrac{2p - 1}{3} - 6 \cdot 4 = 6 \cdot \dfrac{p - 6}{2}$

$2(2p - 1) - 24 = 3(p - 6)$

$4p - 2 - 24 = 3p - 18$

$4p - 26 = 3p - 18$

$p = 8$

29.
$$\frac{m - 3}{8} - \frac{m - 10}{12} = \frac{5}{12}$$

$$24 \cdot \frac{m - 3}{8} - 24 \cdot \frac{m - 10}{12} = 24 \cdot \frac{5}{12}$$

$$3(m - 3) - 2(m - 10) = 10$$

$$3m - 9 - 2m + 20 = 10$$

$$m + 11 = 10$$

$$m = -1$$

31.
$$\frac{h + 7}{2} - \frac{1}{3} = \frac{1}{2} - \frac{h + 9}{9}$$

$$18 \cdot \frac{h + 7}{2} - 18 \cdot \frac{1}{3} = 18 \cdot \frac{1}{2} - 18 \cdot \frac{h + 9}{9}$$

$$9(h + 7) - 6 = 9 - 2(h + 9)$$

$$9h + 63 - 6 = 9 - 2h - 18$$

$$9h + 57 = -2h - 9$$

$$11h = -66$$

$$h = -6$$

33.
$$\frac{x}{x + 4} + 1 = \frac{16}{x + 4}$$

$$(x + 4)\frac{x}{x + 4} + (x + 4)1 = (x + 4)\frac{16}{x + 4}$$

$$x + x + 4 = 16$$

$$2x + 4 = 16$$

$$2x = 12$$

$$x = 6$$

35.
$$\frac{y + 1}{y - 2} = \frac{3}{y - 2} + 5$$

$$(y - 2)\frac{y + 1}{y - 2} = (y - 2)\frac{3}{y - 2} + (y - 2)5$$

$$y + 1 = 3 + 5y - 10$$

$$y + 1 = 5y - 7$$

$$-4y = -8$$

$$y = 2$$

Since $y = 2$ makes two denominators 0, there
is no solution.

37.
$$\frac{5}{x} - \frac{2}{x + 2} = \frac{4}{x}$$

$$x(x + 2)\frac{5}{x} - x(x + 2)\frac{2}{x + 2} = x(x + 2)\frac{4}{x}$$

$$(x + 2)5 - 2x = (x + 2)4$$

$$5x + 10 - 2x = 4x + 8$$

$$3x + 10 = 4x + 8$$

$$2 = x$$

39.
$$\frac{k}{2k + 2} = \frac{2k}{4k + 4} - \frac{k + 3}{k + 1}$$

$$\frac{k}{2(k + 1)} = \frac{2k}{4(k + 1)} - \frac{k + 3}{k + 1}$$

$$\frac{k}{2(k + 1)} = \frac{k}{2(k + 1)} - \frac{k + 3}{k + 1}$$

$$0 = -\frac{k + 3}{k + 1}$$

$$\frac{k + 3}{k + 1} = 0$$

$$(k + 1)\frac{k + 3}{k + 1} = (k + 1)0$$

$$k + 3 = 0$$

$$k = -3$$

41.
$$\frac{2}{t} + \frac{3}{t - 5} = \frac{2(3t + 2)}{t^2 - 5t}$$

$$t(t - 5)\frac{2}{t} + t(t - 5)\frac{3}{t - 5} = t(t - 5)\frac{2(3t + 2)}{t(t - 5)}$$

$$(t - 5)2 + 3t = 2(3t + 2)$$

$$2t - 10 + 3t = 6t + 4$$

$$-10 + 5t = 6t + 4$$

$$-14 = t$$

43.
$$\frac{3}{2p - 2} + \frac{1}{2} = \frac{1}{p^2 - 1}$$

$$\frac{3}{2(p - 1)} + \frac{1}{2} = \frac{1}{(p + 1)(p - 1)}$$

Multiply every term by the LCD $2(p - 1)(p + 1)$.

$$3(p + 1) + (p - 1)(p + 1) = 2$$

$$3p + 3 + p^2 - 1 = 2$$

$$p^2 + 3p = 0$$

$$p(p + 3) = 0$$

$$p = 0 \quad \text{or} \quad p + 3 = 0$$

$$p = -3$$

45.　　$$\frac{1}{m - 3} - \frac{3}{m + 3} = \frac{11}{m^2 - 9}$$

$$\frac{1}{m - 3} - \frac{3}{m + 3} = \frac{11}{(m + 3)(m - 3)}$$

Multiply every term by the LCD $(m - 3)(m + 3)$.

$$m + 3 - 3(m - 3) = 11$$

$$m + 3 - 3m + 9 = 11$$

$$-2m + 12 = 11$$

$$-2m = -1$$

$$m = \frac{1}{2}$$

47.　　$$\frac{2x}{2x - 4} + \frac{8}{x^2 - 4} = \frac{3x}{3x + 6}$$

$$\frac{x}{x - 2} + \frac{8}{(x + 2)(x - 2)} = \frac{x}{x + 2}$$

Multiply every term by the LCD $(x + 2)(x - 2)$.

$$x(x + 2) + 8 = x(x - 2)$$

$$x^2 + 2x + 8 = x^2 - 2x$$

$$8 = -4x$$

$$-2 = x$$

Since $x = -2$ makes two denominators 0, there is no solution.

49.
$$\frac{2}{r} + \frac{8}{r^2 - 16} = \frac{1}{r - 4}$$

$$\frac{2}{r} + \frac{8}{(r + 4)(r - 4)} = \frac{1}{r - 4}$$

Multiply every term by the LCD $r(r + 4)(r - 4)$.

$$2(r + 4)(r - 4) + 8r = r(r + 4)$$

$$2(r^2 - 16) + 8r = r^2 + 4r$$

$$2r^2 - 32 + 8r = r^2 + 4r$$

$$r^2 + 4r - 32 = 0$$

$$(r + 8)(r - 4) = 0$$

$$r + 8 = 0 \quad \text{or} \quad r - 4 = 0$$

$$r = -8 \qquad\qquad r = 4$$

Since $r = 4$ makes two denominators 0, the only solution is $r = -8$.

51.
$$\frac{4y}{y^2 - y - 2} = \frac{7y}{y^2 + y - 6} - \frac{3y - 1}{y^2 + 4y + 3}$$

$$\frac{4y}{(y - 2)(y + 1)} = \frac{7y}{(y + 3)(y - 2)} - \frac{3y - 1}{(y + 3)(y + 1)}$$

Multiply every term by the LCD $(y - 2)(y + 1)(y + 3)$.

$$4y(y + 3) = 7y(y + 1) - (3y - 1)(y - 2)$$

$$4y^2 + 12y = 7y^2 + 7y - (3y^2 - 7y + 2)$$

$$4y^2 + 12y = 7y^2 + 7y - 3y^2 + 7y - 2$$

$$4y^2 + 12y = 4y^2 + 14y - 2$$

$$-2y = -2$$

$$y = 1$$

53.
$$\frac{3}{x^2 + x - 2} - \frac{1}{x^2 - 1} = \frac{7}{2(x^2 + 3x + 2)}$$

$$\frac{3}{(x + 2)(x - 1)} - \frac{1}{(x + 1)(x - 1)} = \frac{7}{2(x + 1)(x + 2)}$$

Multiply by the LCD $2(x + 2)(x - 1)(x + 1)$.

$$3 \cdot 2(x + 1) - 2(x + 2) = 7(x - 1)$$

$$6x + 6 - 2x - 4 = 7x - 7$$

$$4x + 2 = 7x - 7$$

$$9 = 3x$$

$$3 = x$$

55.
$$T = \frac{n}{4} + 40$$

$$4 \cdot T = 4 \cdot \frac{n}{4} + 4 \cdot 40$$

$$4T = n + 160$$

$$4T - 160 = n$$

or $\quad n = 4T - 160$

57.
$$\frac{PV}{T} = \frac{pv}{t}$$

$$T \cdot \frac{PV}{T} = T \cdot \frac{pv}{t}$$

$$PV = \frac{pvT}{t}$$

$$\frac{1}{V} \cdot PV = \frac{1}{V} \cdot \frac{pvT}{t}$$

$$P = \frac{pvT}{tV}$$

59.
$$\frac{PV}{T} = \frac{pv}{t}$$

$$Tt \cdot \frac{PV}{T} = Tt \cdot \frac{pv}{t}$$

$$tPV = pvT$$

$$\frac{tPV}{pv} = \frac{pvT}{pv}$$

$$T = \frac{tPV}{pv}$$

61.
$$I = \frac{E}{R + r}$$

$$(R + r)I = (R + r)\frac{E}{R + r}$$

$$RI + rI = E$$

$$rI = E - RI$$

$$r = \frac{E - RI}{I}$$

63.
$$y = \frac{x + 1}{x - 1}$$

$$(x - 1)y = (x - 1)\frac{x + 1}{x - 1}$$

$$xy - y = x + 1$$

$$xy - x = y + 1$$

$$x(y - 1) = y + 1$$

$$x = \frac{y + 1}{y - 1}$$

65.
$$\frac{1}{z} = \frac{1}{x} + \frac{1}{y}$$

$$xyz \cdot \frac{1}{z} = xyz \cdot \frac{1}{x} + xyz \cdot \frac{1}{y}$$

$$xy = yz + xz$$

$$xy - xz = yz$$

$$x(y - z) = yz$$

$$x = \frac{yz}{y - z}$$

Problem Set 5.8, pp. 259-261

1. x = the number

$$\frac{x}{2} = 3 + \frac{x}{5}$$

$$10 \cdot \frac{x}{2} = 10 \cdot 3 + 10 \cdot \frac{x}{5}$$

$$5x = 30 + 2x$$

$$3x = 30$$

$$x = 10$$

3. x = the number

$$4 + 3 \cdot \frac{1}{x} = 8$$

$$x \cdot 4 + x \cdot \frac{3}{x} = x \cdot 8$$

$$4x + 3 = 8x$$

$$3 = 4x$$

$$\frac{3}{4} = x$$

5. x = first number

3x = second number

$$\frac{1}{x} + \frac{1}{3x} = \frac{2}{3}$$

$$3x \cdot \frac{1}{x} + 3x \cdot \frac{1}{3x} = 3x \cdot \frac{2}{3}$$

$$3 + 1 = 2x$$

$$4 = 2x$$

$$x = 2$$

$$3x = 6$$

7. x = the number

$$\frac{5 + x}{8 + x} = \frac{2}{3}$$

$$3(8 + x)\frac{5 + x}{8 + x} = 3(8 + x)\frac{2}{3}$$

$$3(5 + x) = (8 + x)2$$

$$15 + 3x = 16 + 2x$$

$$x = 1$$

9. x = original salary

$$x - \frac{3}{8}x = 1345$$

$$8 \cdot x - 8 \cdot \frac{3x}{8} = 8 \cdot 1345$$

$$8x - 3x = 10{,}760$$

$$5x = 10{,}760$$

$$x = \$2152$$

11. x = speed of the current

	Distance	Rate	Time
Upstream	4	12 - x	$\frac{4}{12 - x}$
Downstream	5	12 + x	$\frac{5}{12 + x}$

$$\frac{4}{12 - x} = \frac{5}{12 + x}$$

$$(12 - x)(12 + x)\frac{4}{12 - x} = (12 - x)(12 + x)\frac{5}{12 + x}$$

$$4(12 + x) = 5(12 - x)$$

$$48 + 4x = 60 - 5x$$

$$9x = 12$$

$$x = \frac{12}{9} = \frac{4}{3} = 1\frac{1}{3} \text{ mph}$$

13. x = speed of plane in still air

	Distance	Rate	Time
Against wind	30	x - 25	$\dfrac{30}{x - 25}$
With wind	50	x + 25	$\dfrac{50}{x + 25}$

$$\frac{30}{x - 25} = \frac{50}{x + 25}$$

$$(x - 25)(x + 25)\frac{30}{x - 25} = (x - 25)(x + 25)\frac{50}{x + 25}$$

$$30(x + 25) = 50(x - 25)$$

$$30x + 750 = 50x - 1250$$

$$2000 = 20x$$

$$x = 100 \text{ mph}$$

15.

	Distance	Rate	Time
Slow racer	45	x	$\dfrac{45}{x}$
Fast racer	60	x + 10	$\dfrac{60}{x + 10}$

$$\frac{45}{x} = \frac{60}{x + 10}$$

$$x(x + 10)\frac{45}{x} = x(x + 10)\frac{60}{x + 10}$$

$$45(x + 10) = 60x$$

$$45x + 450 = 60x$$

$$450 = 15x$$

$$x = 30 \text{ mph}$$

$$x + 10 = 40 \text{ mph}$$

17.

	Distance	Rate	Time
Drove	110	10x	$\dfrac{110}{10x}$
Walked	2	x	$\dfrac{2}{x}$

$$\frac{110}{10x} + \frac{2}{x} = 3$$

$$\frac{11}{x} + \frac{2}{x} = 3$$

$$x \cdot \frac{11}{x} + x \cdot \frac{2}{x} = x \cdot 3$$

$$11 + 2 = 3x$$

$$13 = 3x$$

$$x = \frac{13}{3} = 4\frac{1}{3} \text{ mph}$$

19.

	Time to do job	Portion of job done in 1 hr
First employee	3	$\dfrac{1}{3}$
Second employee	2	$\dfrac{1}{2}$
Both employees	x	$\dfrac{1}{x}$

$$\frac{1}{3} + \frac{1}{2} = \frac{1}{x}$$

$$6x \cdot \frac{1}{3} + 6x \cdot \frac{1}{2} = 6x \cdot \frac{1}{x}$$

$$2x + 3x = 6$$

$$5x = 6$$

$$x = \frac{6}{5} = 1\frac{1}{5} \text{ hr}$$

21.

	Time to do job	Portion of job done in 1 day
Slow typist	$2x$	$\dfrac{1}{2x}$
Fast typist	x	$\dfrac{1}{x}$
Both typists	4	$\dfrac{1}{4}$

$$\frac{1}{2x} + \frac{1}{x} = \frac{1}{4}$$

$$4x \cdot \frac{1}{2x} + 4x \cdot \frac{1}{x} = 4x \cdot \frac{1}{4}$$

$$2 + 4 = x$$

$$x = 6 \text{ days}$$

$$2x = 12 \text{ days}$$

23.

	Time to do job	Portion of job done in 1 hr
Old press	$x + 5$	$\dfrac{1}{x + 5}$
New press	x	$\dfrac{1}{x}$
Both presses	6	$\dfrac{1}{6}$

$$\frac{1}{x + 5} + \frac{1}{x} = \frac{1}{6}$$

$$6x(x + 5)\frac{1}{x + 5} + 6x(x + 5)\frac{1}{x} = 6x(x + 5)\frac{1}{6}$$

$$6x + 6(x + 5) = x(x + 5)$$

$$12x + 30 = x^2 + 5x$$

$$0 = x^2 - 7x - 30$$

$$0 = (x - 10)(x + 3)$$

$$x - 10 = 0 \quad \text{or} \quad x + 3 = 0$$

$$x = 10 \text{ hr} \qquad \cancel{x = -3}$$

$$x + 5 = 15 \text{ hr}$$

25.

	Time to do job	Portion of job done in 1 hr
Pipe	3	$\frac{1}{3}$
Pump	-5	$-\frac{1}{5}$
Both	x	$\frac{1}{x}$

$$\frac{1}{3} + \left(-\frac{1}{5}\right) = \frac{1}{x}$$

$$15x \cdot \frac{1}{3} - 15x \cdot \frac{1}{5} = 15x \cdot \frac{1}{x}$$

$$5x - 3x = 15$$

$$2x = 15$$

$$x = \frac{15}{2} = 7\frac{1}{2} \text{ hr}$$

27.
$$\frac{1}{R} = \frac{1}{R_1} + \frac{1}{R_2}$$

$$\frac{1}{R} = \frac{1}{6} + \frac{1}{12}$$

$$12R \cdot \frac{1}{R} = 12R \cdot \frac{1}{6} + 12R \cdot \frac{1}{12}$$

$$12 = 2R + R$$

$$12 = 3R$$

$$R = 4 \text{ ohm}$$

29.
$$\frac{1}{R} = \frac{1}{R_1} + \frac{1}{R_2}$$

$$\frac{1}{11} = \frac{1}{R_1} + \frac{1}{16}$$

$$176R_1 \cdot \frac{1}{11} = 176R_1 \cdot \frac{1}{R_1} + 176R_1 \cdot \frac{1}{16}$$

$$16R_1 = 176 + 11R_1$$

$$5R_1 = 176$$

$$R_1 = 35.2 \text{ ohm}$$

Problem Set 5.9, pp. 265-266

1. $\dfrac{1 \text{ in.}}{5 \text{ in.}} = \dfrac{1}{5}$

3. $\dfrac{5 \text{ in.}}{1 \text{ in.}} = \dfrac{5}{1}$

5. $\dfrac{4 \text{ mi}}{6 \text{ mi}} = \dfrac{2}{3}$

7. $\dfrac{5 \text{ days}}{20 \text{ hr}} = \dfrac{120 \text{ hr}}{20 \text{ hr}} = \dfrac{6}{1}$

9. $\dfrac{80 \text{ cents}}{3 \text{ dollars}} = \dfrac{80 \text{ cents}}{300 \text{ cents}} = \dfrac{4}{15}$

11. $\dfrac{7\frac{1}{2} \text{ ft}}{6 \text{ yd}} = \dfrac{7\frac{1}{2} \text{ ft}}{18 \text{ ft}} = 7\frac{1}{2} \div 18 = \dfrac{15}{2} \div \dfrac{18}{1} = \dfrac{15}{2} \cdot \dfrac{1}{18} = \dfrac{5}{12}$

13. $\dfrac{54 \text{ males}}{30 \text{ females}} = \dfrac{9 \text{ males}}{5 \text{ females}}$

15. $\dfrac{162 \text{ cents}}{3 \text{ oz}} = 54¢/\text{oz}$

17. $\dfrac{371 \text{ mi}}{7 \text{ hr}} = 53 \text{ mi/hr}$

19. $\dfrac{2 \text{ oz shampoo}}{2 \text{ oz water}} = \dfrac{1 \text{ oz shampoo}}{1 \text{ oz water}}$

21. $\dfrac{3/4 \text{ lap}}{1/2 \text{ lap}} = \dfrac{3/4}{1/2} = \dfrac{3}{4} \div \dfrac{1}{2} = \dfrac{3}{4} \cdot \dfrac{2}{1} = \dfrac{3}{2}$

23. $\dfrac{x}{16} = \dfrac{3}{4}$

$4x = 48$

$x = 12$

25. $\dfrac{5}{7} = \dfrac{20}{y}$

$5y = 140$

$y = 28$

27. $\dfrac{3}{5} = \dfrac{m}{4}$

$5m = 12$

$m = \dfrac{12}{5}$

29. $\dfrac{12}{r} = \dfrac{8}{2}$

$\dfrac{12}{r} = \dfrac{4}{1}$

$4r = 12$

$r = 3$

31. $\dfrac{x}{3} = \dfrac{3}{5}$

$5x = 9$

$x = \dfrac{9}{5}$

33. $\dfrac{8}{x} = \dfrac{6\frac{2}{3}}{35}$

$\dfrac{8}{x} = \dfrac{20/3}{35}$

$\dfrac{20}{3}x = 280$

$\dfrac{3}{20}\cdot\dfrac{20}{3}x = \dfrac{3}{20}\cdot 280$

$x = 42$

35. $\dfrac{1/2}{1/3} = \dfrac{3/4}{z}$

$\dfrac{1}{2}z = \dfrac{1}{3}\cdot\dfrac{3}{4}$

$\dfrac{1}{2}z = \dfrac{1}{4}$

$2\cdot\dfrac{1}{2}z = 2\cdot\dfrac{1}{4}$

$z = \dfrac{1}{2}$

37. $\dfrac{3.6}{2.4} = \dfrac{t}{6.5}$

$2.4t = (3.6)(6.5)$

$2.4t = 23.4$

$t = \dfrac{23.4}{2.4}$

$t = 9.75$

39. $\dfrac{5}{8} = \dfrac{p}{p + 1}$

$5(p + 1) = 8p$

$5p + 5 = 8p$

$5 = 3p$

$p = \dfrac{5}{3}$

41. $\dfrac{k - 2}{k + 4} = \dfrac{4}{k}$

$k(k - 2) = 4(k + 4)$

$k^2 - 2k = 4k + 16$

$k^2 - 6k - 16 = 0$

$(k - 8)(k + 2) = 0$

$k - 8 = 0 \quad \text{or} \quad k + 2 = 0$

$k = 8 \qquad\qquad k = -2$

43. x = height of flagpole

Height of flagpole $\longrightarrow \dfrac{x}{28} = \dfrac{5}{4} \longleftarrow$ Height of tree
Shadow of flagpole $\longrightarrow$ $\longleftarrow$ Shadow of tree

$$4x = 140$$

$$x = 35 \text{ ft}$$

45. x = number of males

Males $\longrightarrow \dfrac{x}{910} = \dfrac{9}{7} \longleftarrow$ Males
Females $\longrightarrow$ $\longleftarrow$ Females

$$7x = 8190$$

$$x = 1170 \text{ males}$$

47. x = length of enlargement

Length of photo $\longrightarrow \dfrac{5}{4} = \dfrac{x}{10} \longleftarrow$ Length of enlargement
Width of photo $\longrightarrow$ $\longleftarrow$ Width of enlargement

$$4x = 50$$

$$x = \frac{50}{4} = \frac{25}{2} = 12.5 \text{ in.}$$

49. x = number of defective bulbs in shipment

Total bulbs in shipment $\longrightarrow \dfrac{2250}{x} = \dfrac{1000}{8} \longleftarrow$ Total bulbs in sample
Defective bulbs in shipment $\longrightarrow$ $\longleftarrow$ Defective bulbs in sample

$$1000x = 18{,}000$$

$$x = 18 \text{ bulbs}$$

51. x = lawn covered by $5\frac{1}{2}$ bags

Bags of fertilizer ⟶ $\dfrac{3\frac{1}{2}}{2800}$ = $\dfrac{5\frac{1}{2}}{x}$ ⟵ Bags of fertilizer
Lawn covered ⟶ ⟵ Lawn covered

$$\frac{7/2}{2800} = \frac{11/2}{x}$$

$$\frac{7}{2}x = 2800 \cdot \frac{11}{2}$$

$$\frac{7}{2}x = 15{,}400$$

$$\frac{2}{7} \cdot \frac{7}{2}x = \frac{2}{7} \cdot 15{,}400$$

$$x = 4400 \text{ sq ft}$$

53. x = time to type 99 pages

Pages ⟶ $\dfrac{12}{1\frac{20}{60}}$ = $\dfrac{99}{x}$ ⟵ Pages
Time ⟶ ⟵ Time

$$\frac{12}{1\frac{1}{3}} = \frac{99}{x}$$

$$\frac{12}{4/3} = \frac{99}{x}$$

$$12x = \frac{4}{3} \cdot 99$$

$$12x = 132$$

$$x = 11 \text{ hr}$$

55. x = number of deer in the park

Total deer in park $\longrightarrow$ $\dfrac{x}{62}$ = $\dfrac{45}{6}$ $\longleftarrow$ Total deer in sample
Tagged deer in park $\longrightarrow$ $\longleftarrow$ Tagged deer in sample

$$6x = 2790$$

$$x = 465 \text{ deer}$$

57.   x = amount first person receives

6120 - x = amount second person receives

$$\frac{x}{6120 - x} = \frac{5}{7}$$

$$7x = 5(6120 - x)$$

$$7x = 30,600 - 5x$$

$$12x = 30,600$$

$$x = \$2550$$

$$6120 - x = \$3570$$

59. $\dfrac{BD}{AD} = \dfrac{BE}{CE}$

$$\frac{x}{14} = \frac{42 - x - z}{84}$$

$$\frac{x}{14} = \frac{42 - x - 7}{84}$$

$$\frac{x}{14} = \frac{35 - x}{84}$$

$$84x = 14(35 - x)$$

$$84x = 490 - 14x$$

$$98x = 490$$

$$x = 5 \text{ in.}$$

NOTES

CHAPTER 6

LINEAR EQUATIONS AND INEQUALITIES IN
TWO VARIABLES

Problem Set 6.1, pp. 275-278

1. Quadrant I

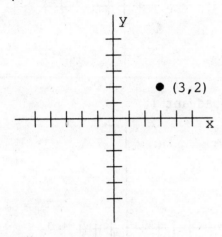

3. Quadrant IV

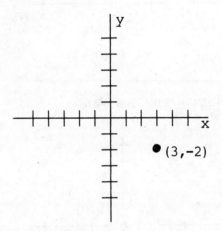

5. Quadrant II

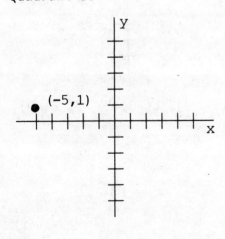

7. Quadrant III

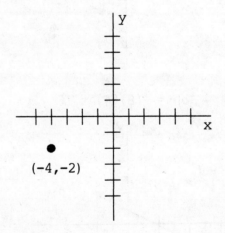

9. **Not in any quadrant; on the negative part of the _y_-axis**

11. **Not in any quadrant; on the positive part of the _x_-axis**

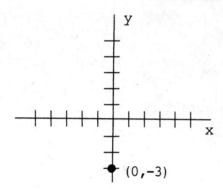

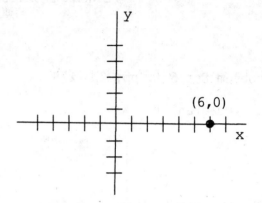

13. **Not in any quadrant**

15. **Quadrant IV**

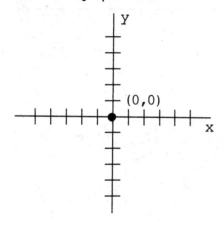

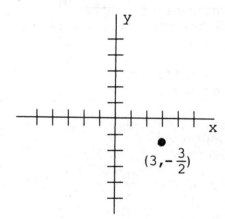

17. $A = \frac{1}{2}bh = \frac{1}{2}(8)(4) = 16$

19. A(5, 2), C(-3, 2), E(-5, 0), G(0, 0)

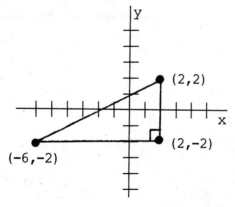

21. and 23.

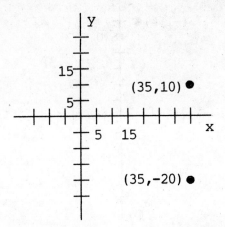

25. and 27.

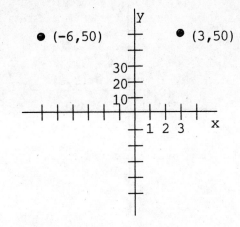

29. Quadrant IV

31. Quadrant I

33. Quadrant I, Quadrant II,
     or the positive part of
     the y-axis.

35. Both coordinates are negative.

37. The x-coordinate is negative
     and the y-coordinate is positive.

39. The x-coordinate is positive
     and the y-coordinate is 0.

41. (a, b) = (+, -)

     Quadrant IV

43. (a, -b) = (+, +)

     Quadrant I

45. (a, b - a) = (+, -)

     Quadrant IV

47.

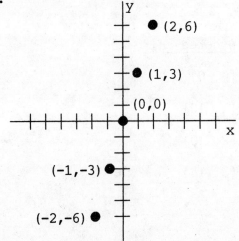

49.

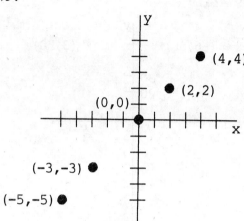

51.

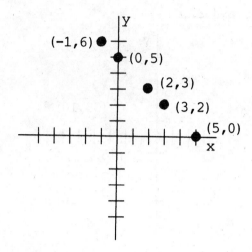

53.

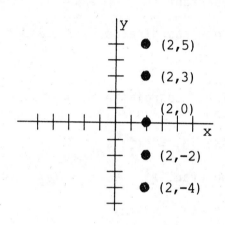

55.

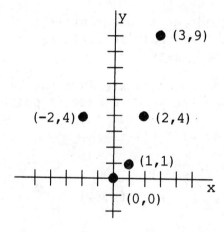

57.

Ordered Pair	Interpretation
(0, 0)	It costs $0 to park 0 hr.
(1, 10)	It costs $10 to park 1 hr.
(2, 10)	It costs $10 to park 2 hr.
(3, 15)	It costs $15 to park 3 hr.

59. <u>Ordered Pair</u>   <u>Interpretation</u>

(0, 70)   0 min after mixing, the temperature is 70°.

(5, 20)   5 min after mixing, the temperature is 20°.

(20, -20)   20 min after mixing, the temperature is -20°.

61. (1, 4), (2, 3), (3, 2), (4, 1)

For example, the ordered pair (1, 4) represents a 1 on the red die
and a 4 on the green die.

63. (1, 6), (2, 5), (3, 4), (4, 3), (5, 2), (6, 1), (2, 6), (3, 5),

(4, 4), (5, 3), (6, 2)

Problem Set 6.2, pp. 286-288

1. $3x + 2y = 8$

(a) $3(0) + 2(4) = 8$

$0 + 8 = 8$

$8 = 8$ True

(0, 4) is a solution.

(b) $3(2) + 2(-1) = 8$

$6 - 2 = 8$

$4 = 8$ False

(2, -1) is not a solution.

(c) $3(-2) + 2(7) = 8$

$-6 + 14 = 8$

$8 = 8$ True

(-2, 7) is a solution.

3. $y = -4x + 7$

(a) $5 = -4(3) + 7$

$5 = -12 + 7$

$5 = -5$ False

(3, 5) is not a solution.

(b) $5 = -4(\frac{1}{2}) + 7$

$5 = -2 + 7$

$5 = 5$ True

($\frac{1}{2}$, 5) is a solution.

(c) $23 = -4(-4) + 7$

$23 = 16 + 7$

$23 = 23$ True

(-4, 23) is a solution.

5. y = 4

    (a) 0 = 4 False

    (0, 0) is not a solution.

    (b) 4 = 4 True

    (5, 4) is a solution.

    (c) 5 = 4 False

    (4, 5) is not a solution.

7. x = -6

    (a) 1 = -6 False

    (1, -6) is not a solution.

    (b) -6 = -6 True

    (-6, 0) is a solution.

    (c) -6 = -6 True

    (-6, 10) is a solution.

9. y = x + 1

x	y
0	1
1	2
2	3
-1	0

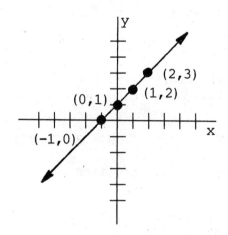

11. x + 2y = 6

x	y
0	3
6	0
2	2

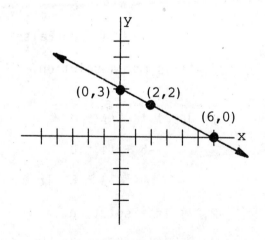

13. $y = \frac{3}{4}x$

x	y
0	0
4	3
-4	-3

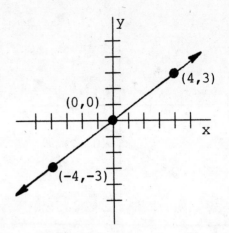

15. $y = 2x - 4$

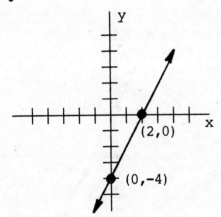

17. $x + y = 4$

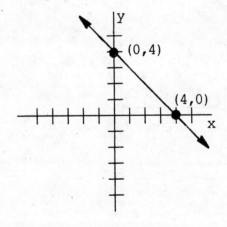

19. $4x + y = 4$

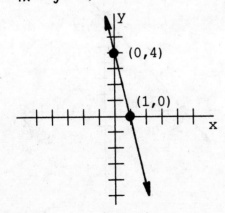

21. $3x + 5y = -15$

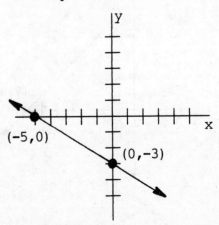

23. 2x - 3y = 6

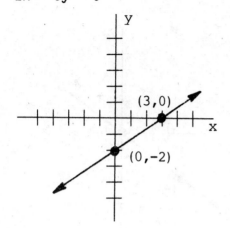

25. y = x

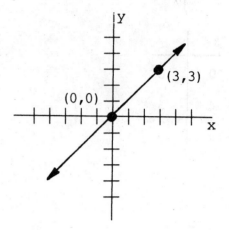

27. y = 3x

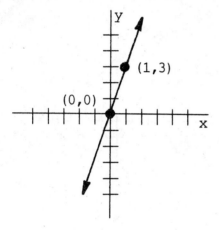

29. y = -2x

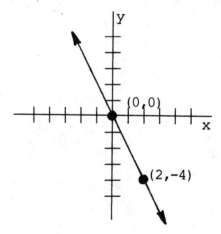

31. y = -x + 7

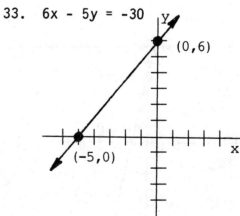

33. 6x - 5y = -30

35. $-2x + 7y = 21$

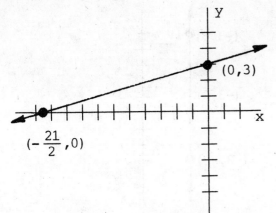

(0,3)

$(-\frac{21}{2},0)$

37. $-9x - 18y = 24$

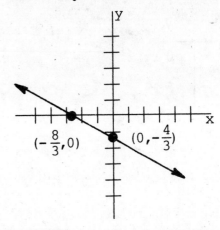

$(-\frac{8}{3},0)$ $(0,-\frac{4}{3})$

39. $y = \frac{1}{2}x$

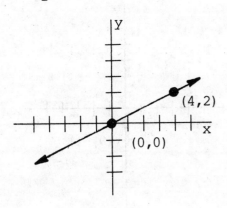

(4,2)

(0,0)

41. $y = \frac{2}{3}x + 6$

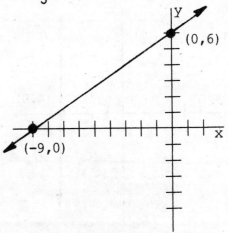

(0,6)

(-9,0)

43. $y = 2$

x	y
0	2
4	2
-4	2

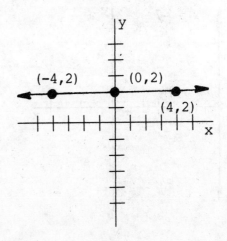

(-4,2)   (0,2)

(4,2)

45. x = -3

x	y
-3	0
-3	5
-3	-3

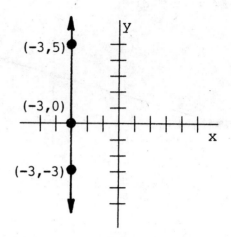

47. x = 4

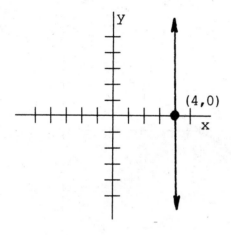

49. y = -1

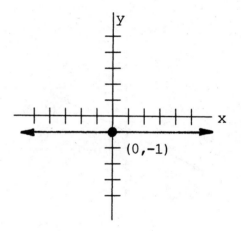

51. x + 7 = 0

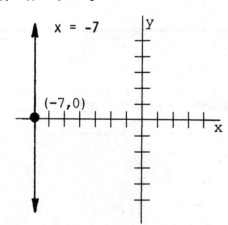

53. y - 3 = 0

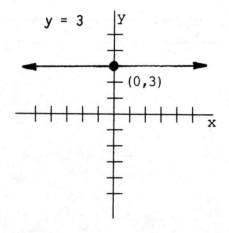

55. 2x - 5 = 0

2x = 5

$x = \frac{5}{2}$

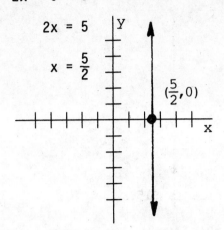

$(\frac{5}{2}, 0)$

57. y = 0

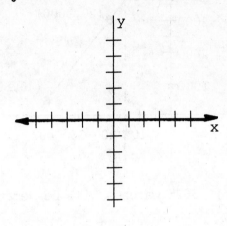

59. y = 2x + 5

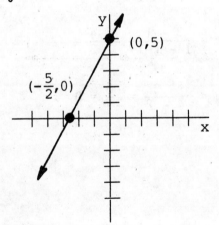

(0,5)

$(-\frac{5}{2}, 0)$

61. x + 6y = 9

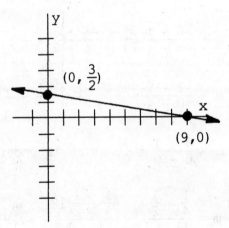

$(0, \frac{3}{2})$

(9,0)

63. y = 5

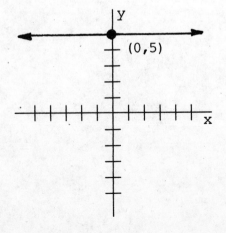

(0,5)

65. (a) $100

(b) $150

(c) 30 calculators

67. (a)

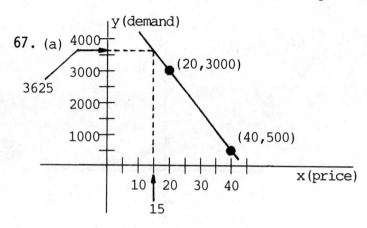

(b)   3625 watches will be demanded if the price is $15.

Problem Set 6.3, pp. 294-296

1. $m = \dfrac{y_2 - y_1}{x_2 - x_1} = \dfrac{6 - 1}{8 - 2} = \dfrac{5}{6}$

3. $m = \dfrac{y_2 - y_1}{x_2 - x_1} = \dfrac{3 - 1}{6 - 0} = \dfrac{2}{6} = \dfrac{1}{3}$

5. $m = \dfrac{-6 - (-3)}{5 - (-2)} = \dfrac{-6 + 3}{5 + 2} = -\dfrac{3}{7}$

7. $m = \dfrac{-3 - 1}{1 - (-1)} = \dfrac{-4}{2} = -2$

9. $m = \dfrac{-2 - (-11)}{0 - 4} = \dfrac{-2 + 11}{-4} = -\dfrac{9}{4}$

11. $m = \dfrac{3 - 3}{-4 - 2} = \dfrac{0}{-6} = 0$

13. $m = \dfrac{4 - 1}{6 - 6} = \dfrac{3}{0}$, which is undefined.

15. $m = \dfrac{\frac{1}{4} - \frac{1}{2}}{\frac{11}{12} - \frac{1}{3}} = \dfrac{(\frac{1}{4} - \frac{1}{2})12}{(\frac{11}{12} - \frac{1}{3})12} = \dfrac{3 - 6}{11 - 4} = -\dfrac{3}{7}$

17. $m = \dfrac{5.8 - 1.9}{2.7 - 8.4} = \dfrac{3.9}{-5.7} = -\dfrac{3.9}{5.7} = -\dfrac{39}{57} = -\dfrac{13}{19}$

19.

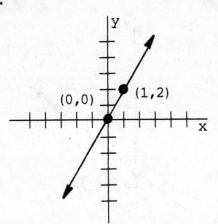

21.

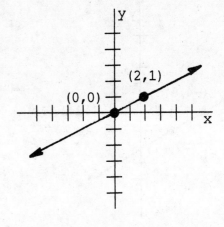

23.

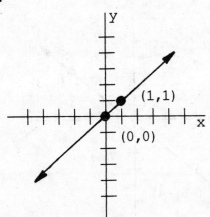

25.

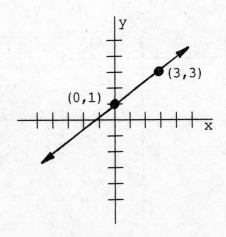

27.

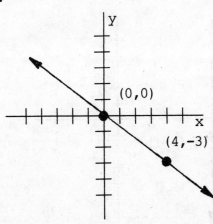

29.

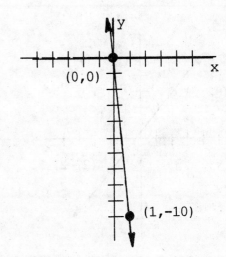

31.

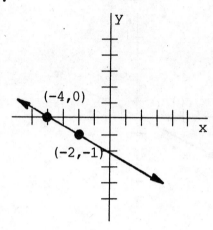

33.

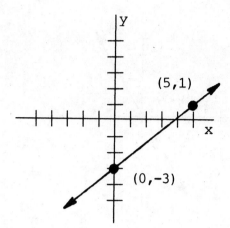

35.

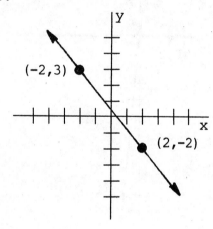

37.

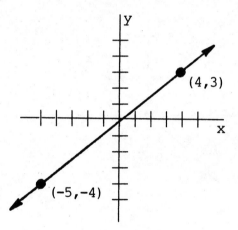

39.

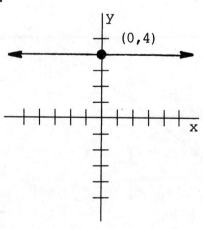

41.

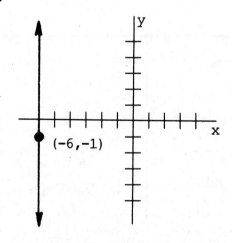

43. $m = \dfrac{6}{2} = 3$

45. $m = \dfrac{-3}{6} = -\dfrac{1}{2}$

47. $m = 0$

49. $m_{AB} = \dfrac{-1 - 1}{2 - 6} = \dfrac{-2}{-4} = \dfrac{1}{2}$

$m_{BC} = \dfrac{-3 - (-1)}{-2 - 2} = \dfrac{-2}{-4} = \dfrac{1}{2}$

Since the line through A and B has the same slope as the line through B and C, and since the lines share the common point B, the lines are the same line.

51. Pitch $= m = \dfrac{rise}{run} = \dfrac{7}{12}$

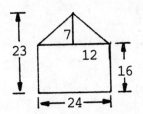

53. By the Pythagorean theorem,

$c^2 = 5^2 + 12^2$

$c^2 = 25 + 144$

$c^2 = 169$

$c = 13.$

Since the triangles are similar, we can set up the following proportion.

$\dfrac{x}{3718} = \dfrac{5}{c}$

$\dfrac{x}{3718} = \dfrac{5}{13}$

$x = \dfrac{5}{13}(3718)$

$x = 1430$ ft

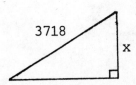

Problem Set 6.4, pp. 303-304

1. $y - y_1 = m(x - x_1)$

   $y - 5 = 3(x - 2)$

   $y - 5 = 3x - 6$

   $-3x + y = -1$

   $-1(-3x + y) = -1(-1)$

   $3x - y = 1$

3. $y - y_1 = m(x - x_1)$

   $y - 7 = 4(x - (-1))$

   $y - 7 = 4(x + 1)$

   $y - 7 = 4x + 4$

   $-4x + y = 11$

   $-1(-4x + y) = -1(11)$

   $4x - y = -11$

5. $y - y_1 = m(x - x_1)$

   $y - (-4) = \frac{1}{2}(x - 6)$

   $y + 4 = \frac{1}{2}(x - 6)$

   $2y + 8 = x - 6$

   $-x + 2y = -14$

   $x - 2y = 14$

7. $y - y_1 = m(x - x_1)$

   $y - (-2) = \frac{7}{3}(x - (-3))$

   $y + 2 = \frac{7}{3}(x + 3)$

   $3y + 6 = 7x + 21$

   $-7x + 3y = 15$

   $7x - 3y = -15$

9. $y - y_1 = m(x - x_1)$

   $y - 9 = -1(x - 0)$

   $y - 9 = -x$

   $x + y = 9$

11. $y - y_1 = m(x - x_1)$

   $y - 0 = -\frac{2}{5}(x - (-8))$

   $y = -\frac{2}{5}(x + 8)$

   $5y = -2(x + 8)$

   $5y = -2x - 16$

   $2x + 5y = -16$

13. $m = \dfrac{5 - 3}{5 - 4} = \dfrac{2}{1} = 2$

$y - 3 = 2(x - 4)$

$y - 3 = 2x - 8$

$-2x + y = -5$

$2x - y = 5$

15. $m = \dfrac{6 - 4}{1 - (-3)} = \dfrac{2}{4} = \dfrac{1}{2}$

$y - 6 = \dfrac{1}{2}(x - 1)$

$2y - 12 = x - 1$

$-x + 2y = 11$

$x - 2y = -11$

17. $m = \dfrac{4 - (-3)}{-5 - 2} = \dfrac{7}{-7} = -1$

$y - (-3) = -1(x - 2)$

$y + 3 = -x + 2$

$x + y = -1$

19. $m = \dfrac{-9 - 6}{1 - (-5)} = \dfrac{-15}{6} = -\dfrac{5}{2}$

$y - 6 = -\dfrac{5}{2}(x - (-5))$

$y - 6 = -\dfrac{5}{2}(x + 5)$

$2y - 12 = -5(x + 5)$

$2y - 12 = -5x - 25$

$5x + 2y = -13$

21. $m = \dfrac{-8 - 0}{-10 - 0} = \dfrac{-8}{-10} = \dfrac{4}{5}$

$y - 0 = \dfrac{4}{5}(x - 0)$

$y = \dfrac{4}{5}x$

$5y = 4x$

$-4x + 5y = 0$

$4x - 5y = 0$

23. $m = \dfrac{7 - 7}{3 - 6} = \dfrac{0}{-3} = 0$

$y - 7 = 0(x - 6)$

$y - 7 = 0$

$y = 7$

25. $y = mx + b$

    $y = 5x + 3$

27. $y = mx + b$

    $y = 1x + (-12)$

    $y = x - 12$

29. $y = mx + b$

    $y = -\frac{1}{2}x + 1$

31. $y = mx + b$

    $y = \frac{4}{7}x + 0$

    $y = \frac{4}{7}x$

33. $y = mx + b$

    $y = -8x + (-\frac{2}{3})$

    $y = -8x - \frac{2}{3}$

35. $y = mx + b$

    $y = 0x + (-5)$

    $y = -5$

37. $y = 4x + 8$

    $m = 4, b = 8$

39. $5x + 3y = 12$

    $3y = -5x + 12$

    $y = -\frac{5}{3}x + 4$

    $m = -\frac{5}{3}, b = 4$

41. $4x - 6y = 0$

    $-6y = -4x$

    $y = \frac{-4}{-6}x$

    $y = \frac{2}{3}x + 0$

    $m = \frac{2}{3}, b = 0$

43. $y = 9$

    $y = 0x + 9$

    $m = 0, b = 9$

45. y = x + 3

m = 1, b = 3

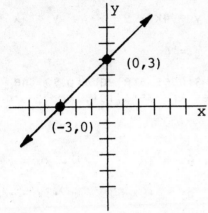

47. y = -3x

m = -3, b = 0

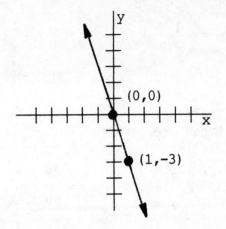

49. 3x - y = 6

-y = -3x + 6

y = 3x - 6

m = 3, b = -6

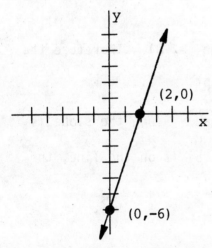

51. 2x - 5y = 10

-5y = -2x + 10

$y = \frac{2}{5}x - 2$

$m = \frac{2}{5}$, b = -2

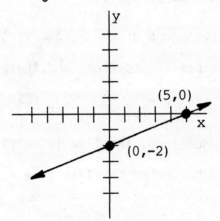

53. y = 6x - 1   y = 6x + 1

m = 6   m = 6

The slopes are equal, so the lines are parallel.

55. y = -5x $y = \frac{x}{5}$

m = -5 $m = \frac{1}{5}$

Since $-5(\frac{1}{5}) = -1$, the lines are perpendicular.

57. $y = -\frac{x}{3} + \frac{2}{3}$ $x + 3y = 0$

$y = -\frac{1}{3}x + \frac{2}{3}$ $3y = -x$

$m = -\frac{1}{3}$ $y = -\frac{1}{3}x$

    $m = -\frac{1}{3}$

The slopes are equal, so the lines are parallel.

59. $4x - y = 9$ $-8x + 2y = 10$

$-y = -4x + 9$ $2y = 8x + 10$

$y = 4x - 9$ $y = 4x + 5$

$m = 4$ $m = 4$

The slopes are equal, so the lines are parallel.

61. $-2x + y = 0$ $x - 2y = 5$

$y = 2x$ $-2y = -x + 5$

$m = 2$ $y = \frac{1}{2}x - \frac{5}{2}$

    $m = \frac{1}{2}$

The slopes are not equal and their product is not -1, so the lines are neither parallel nor perpendicular.

63. $9x + 6y = 18$ $2x - 3y = 6$

$6y = -9x + 18$ $-3y = -2x + 6$

$y = -\frac{3}{2}x + 3$ $y = \frac{2}{3}x - 2$

$m = -\frac{3}{2}$ $m = \frac{2}{3}$

Since $(-\frac{3}{2})\frac{2}{3} = -1$, the lines are perpendicular.

65. The graph of $x = 3$ is a vertical line through (3, 0). Therefore the slope is undefined and there is no y-intercept.

67. $m = \frac{5 - 3}{-1 - (-1)} = \frac{2}{-1 + 1} = \frac{2}{0}$, which is undefined. Since the slope is undefined, the line is vertical. Since (-1, 5) is on the line, the equation of the line is $x = -1$.

69. $y = 4x - 5$

$m = 4$

The desired line is parallel to $y = 4x - 5$, so its slope is also 4.

$y - y_1 = m(x - x_1)$

$y - 3 = 4(x - 1)$

$y - 3 = 4x - 4$

$-4x + y = -1$

$4x - y = 1$

71. $3x - y = 7$

$-y = -3x + 7$

$y = 3x - 7$

$m = 3$

The desired line is perpendicular to $3x - y = 7$, so its slope is $-\frac{1}{3}$.

$y - y_1 = m(x - x_1)$

$y - (-4) = -\frac{1}{3}(x - 2)$

$y + 4 = -\frac{1}{3}(x - 2)$

$3y + 12 = -(x - 2)$

$3y + 12 = -x + 2$

$x + 3y = -10$

73. (a) Two points on the line are (30, 5000) and (35, 4000).

$m = \dfrac{5000 - 4000}{30 - 35} = \dfrac{1000}{-5} = -200$

$y - y_1 = m(x - x_1)$

$y - 5000 = -200(x - 30)$

$y - 5000 = -200x + 6000$

$y = -200x + 11,000$

(b) $y = -200x + 11,000$

$y = -200(20) + 11,000$

$y = -4000 + 11,000$

$y = 7000$ sunglasses

Problem Set 6.5, pp. 309-310

1.   x < 5

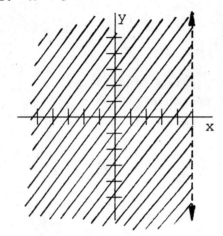

3.   y ≥ -3

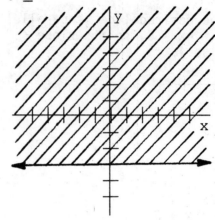

5.   x + y < 4

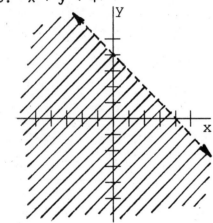

7.   x + y > 4

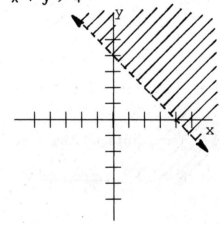

9.   4x - 5y ≤ 20

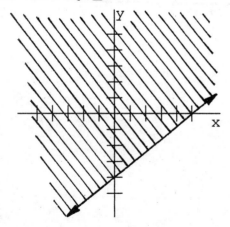

11.   y ≥ 2x

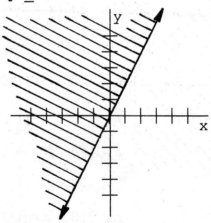

13. x > 1

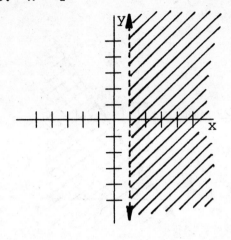

15. y ≤ 3

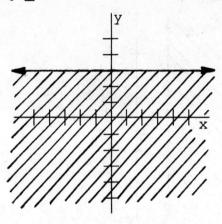

17. y < x + 2

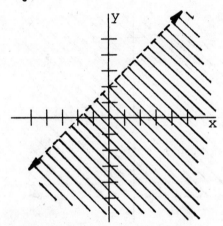

19. y ≤ 2x - 4

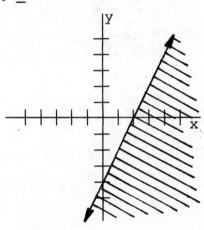

21. x + y < 1

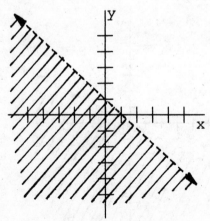

23. x + 2y > 4

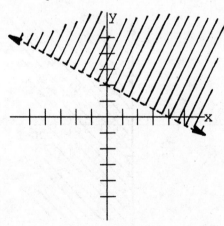

25. $x \leq -2$

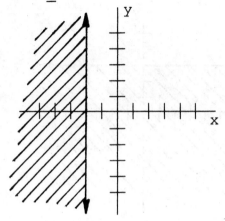

27. $3x - 2y \geq 6$

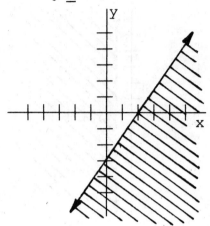

29. $y \geq 3x$

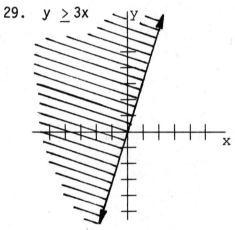

31. $y > -5$

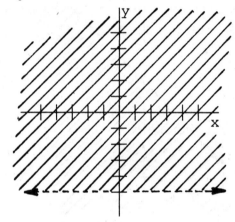

33. $x \geq 0$

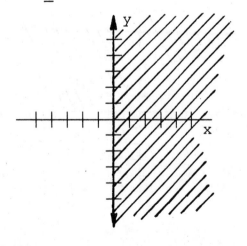

35. $x - 2y > 0$

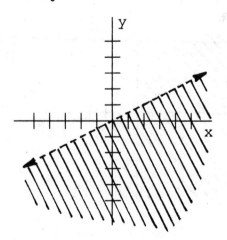

37. $2x + 8y \leq 16$

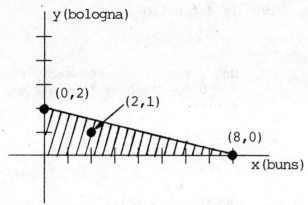

39. $3x + 5y \geq 600$

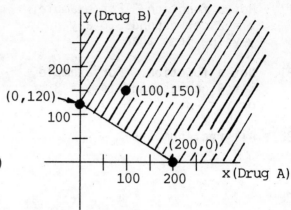

Problem Set 6.6, pp. 316-318

1. Domain = $\left\{2, -6, 0, 9\right\}$

   Range = $\left\{5, 4, 3\right\}$

   Is a function.

3. Domain = $\left\{3, 1\right\}$

   Range = $\left\{9, 7, -8\right\}$

   Not a function since 3 corresponds to 9 and -8.

5. Domain = $\left\{1, 2, 3\right\}$

   Range = $\left\{1, 2, 3\right\}$

   Is a function.

7. Domain = $\left\{-5, -6, -7\right\}$

   Range = $\left\{0\right\}$

   Is a function.

9. Domain = $\left\{1, 2, 3\right\}$

   Range = $\left\{100\right\}$

   Is a function.

11. Domain = $\left\{7, -2, 11\right\}$

   Range = $\left\{-8, 6, 9\right\}$

   Not a function since 11 corresponds to 6 and 9.

13. Is a function.

15. Is a function.

17. Not a function. For example, $x = 2$ corresponds to any y-value less than 6.

19. Is a function.

21. Not a function. For example, $x = 9$ corresponds to $y = 3$ or $y = -3$.

23. Not a function. For example, $x = 0$ corresponds to $y = 2$ or $y = -2$.

25. All real numbers

27. $x - 3 = 0$

   $x = 3$

   Domain is $x \neq 3$.

29. $x^2 - 1 = 0$

   $(x + 1)(x - 1) = 0$

   $x + 1 = 0$ or $x - 1 = 0$

   $x = -1$ $x = 1$

   Domain is $x \neq -1, 1$.

31. $x^2 + 8x = 0$

   $x(x + 8) = 0$

   $x = 0$ or $x + 8 = 0$

   $x = -8$

   Domain is $x \neq 0, -8$.

33. All real numbers

35. All real numbers

37. Since $(x - 1)^2 \geq 0$, the range is $y \geq 0$.

39. Since $x^2 \geq 0$, the range is $y \geq 1$.

41. Yes

43. No, it fails the vertical line test.

45. Yes

47. No, it fails the vertical line test.

49. $f(x) = x + 1$

    (a) $f(2) = 2 + 1 = 3$

    (b) $f(0) = 0 + 1 = 1$

    (c) $f(-1) = -1 + 1 = 0$

51. $f(x) = -6x$

    (a) $f(2) = -6(2) = -12$

    (b) $f(0) = -6(0) = 0$

    (c) $f(-1) = -6(-1) = 6$

53. $f(x) = 4x + 3$

    (a) $f(2) = 4(2) + 3 = 8 + 3 = 11$

    (b) $f(0) = 4(0) + 3 = 0 + 3 = 3$

    (c) $f(-1) = 4(-1) + 3 = -4 + 3 = -1$

55. $f(x) = |x - 2|$

    (a) $f(2) = |2 - 2| = |0| = 0$

    (b) $f(0) = |0 - 2| = |-2| = 2$

    (c) $f(-1) = |-1 - 2| = |-3| = 3$

57. $g(x) = x^2 + x$

    (a) $g(1) = 1^2 + 1 = 1 + 1 = 2$

    (b) $g(0) = 0^2 + 0 = 0 + 0 = 0$

    (c) $g(-3) = (-3)^2 + (-3) = 9 - 3 = 6$

59. $g(x) = 2x^2 - 5x + 9$

    (a) $g(1) = 2(1)^2 - 5(1) + 9 = 2 - 5 + 9 = 6$

    (b) $g(0) = 2(0)^2 - 5(0) + 9 = 0 - 0 + 9 = 9$

    (c) $g(-3) = 2(-3)^2 - 5(-3) + 9 = 2(9) + 15 + 9 = 42$

61. $g(x) = -x^2 + 6x + 4$

    (a) $g(1) = -1^2 + 6(1) + 4 = -1 + 6 + 4 = 9$

    (b) $g(0) = -0^2 + 6(0) + 4 = 0 + 0 + 4 = 4$

    (c) $g(-3) = -(-3)^2 + 6(-3) + 4 = -9 - 18 + 4 = -23$

63. $g(x) = x^3 - 5x^2 + 7x - 4$

   (a) $g(1) = 1^3 - 5(1)^2 + 7(1) - 4 = 1 - 5 + 7 - 4 = -1$

   (b) $g(0) = 0^3 - 5(0)^2 + 7(0) - 4 = 0 - 0 + 0 - 4 = -4$

   (c) $g(-3) = (-3)^3 - 5(-3)^2 + 7(-3) - 4 = -27 - 5(9) - 21 - 4$

$$= -27 - 45 - 21 - 4$$

$$= -97$$

65.   f(Kim) = age of Kim

         = 18 yr

67.   g(Kim) = weight of Kim

         = 100 lb

69.   h(Cindy) = height of Cindy

         = 64 in.

71.   f(Sean) = age of Sean

         = 18 yr

73.   Yes, because each person in the domain has exactly one social
      security number.

75.   No, because some people in the domain have more than one child.

77. $A = x^2$, domain is $x > 0$, range is $A > 0$.

79. $R = 220 - a$, domain is $18 \leq a \leq 55$, range is $165 \leq R \leq 202$.

81. $E(x) = 100 + 0.25x$

   (a) $E(0) = 100 + 0.25(0) = 100 + 0 = 100$ means he earns $100 if
         his sales are $0.

   (b) $E(700) = 100 + 0.25(700) = 100 + 175 = 275$ means he earns
         $275 is his sales are $700.

Chapter 7

LINEAR SYSTEMS

Problem Set 7.1, pp. 330-331

1. $x + y = 15$ $x - y = 7$

      $11 + 4 = 15$ $11 - 4 = 7$

         $15 = 15$ True $7 = 7$ True

    $(11, 4)$ is a solution.

3. $y = 3x - 2$ $y = -4x + 5$

      $1 = 3(-1) - 2$ $1 = -4(-1) + 5$

      $1 = -5$ False $1 = 9$ False

    $(-1, 1)$ is not a solution.

5. $y = 2x + 4$ $6x - 5y = 8$

      $-2 = 2(-3) + 4$ $6(-3) - 5(-2) = 8$

      $-2 = -6 + 4$ $-18 + 10 = 8$

      $-2 = -2$ True $-8 = 8$ False

    $(-3, -2)$ is not a solution.

7. $x - 9y = 6$ $y = \frac{1}{3}x - 2$

      $6 - 9(0) = 6$ $0 = \frac{1}{3}(6) - 2$

           $6 = 6$ True $0 = 0$ True

    $(6, 0)$ is a solution.

9. $x - y = 9$ $y + 4 = 0$

     $5 - (-4) = 9$ $-4 + 4 = 0$

             $9 = 9$ True $0 = 0$ True

(5, -4) is a solution.

11. $8x - 6y = -5$ $x = \frac{3}{2}y - 1$

     $8(-\frac{1}{4}) - 6(\frac{1}{2}) = -5$ $-\frac{1}{4} = \frac{3}{2}(\frac{1}{2}) - 1$

             $-2 - 3 = -5$ $-\frac{1}{4} = \frac{3}{4} - 1$

                    $-5 = -5$ True $-\frac{1}{4} = -\frac{1}{4}$ True

$(-\frac{1}{4}, \frac{1}{2})$ is a solution.

13. Independent & consistent

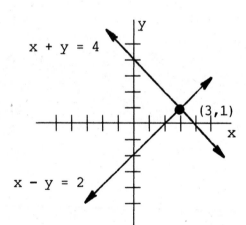

15. Independent & consistent

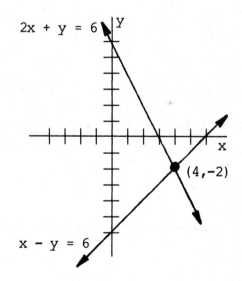

17. Independent & consistent

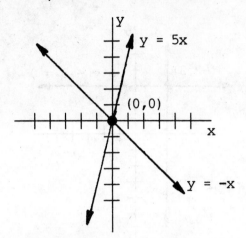

19. Independent & consistent

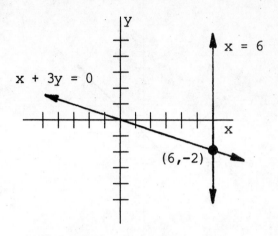

21. Independent & consistent

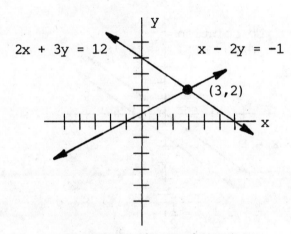

23. Independent & inconsistent

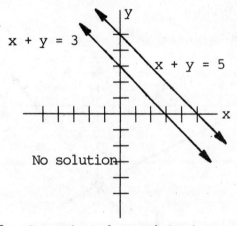

25. Independent & consistent

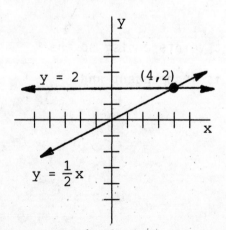

27. Dependent & consistent

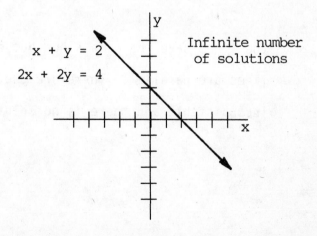

29.   Independent & consistent

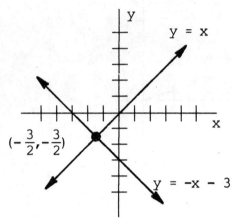

31.   Independent & consistent

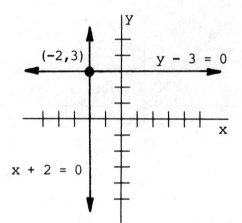

33.   Independent & consistent

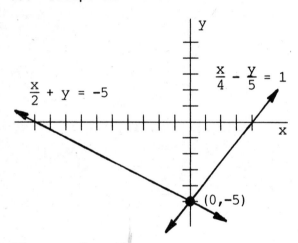

35.   Independent & inconsistent

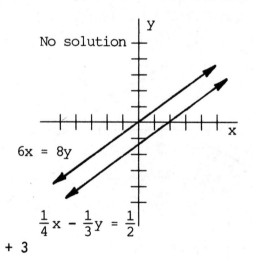

37. $y = 2x - 4$ $y = 2x + 3$

   $m = 2, b = -4$ $m = 2, b = 3$

The lines have the same slope but different y-intercepts, so the
lines are parallel. Therefore the system is independent and
inconsistent, and there is no solution.

39. $5y = 7x$ $5x = 7y$

   $y = \frac{7}{5}x$ $y = \frac{5}{7}x$

   $m = \frac{7}{5},\ b = 0$ $m = \frac{5}{7},\ b = 0$

The lines have different slopes, so the lines intersect at one point. Therefore the system is independent and consistent, and there is one solution.

41. $8x - 2y = 6$ $-4x + y = -3$

      $-2y = -8x + 6$ $y = 4x - 3$

       $y = 4x - 3$ $m = 4,\ b = -3$

   $m = 4,\ b = -3$

The lines have the same slope and the same y-intercept, so the lines are the same line. Therefore the system is dependent and consistent, and there is an infinite number of solutions.

43. $3x - y = 9$ $-x + 4y = 8$

      $-y = -3x + 9$ $4y = x + 8$

       $y = 3x - 9$ $y = \frac{1}{4}x + 2$

   $m = 3,\ b = -9$

                                  $m = \frac{1}{4},\ b = 2$

The lines have different slopes, so the lines intersect at one point. Therefore the system is independent and consistent, and there is one solution.

45. $4x + 6y = -2$ $\qquad\qquad$ $\frac{2}{3}x + y = \frac{1}{3}$

$\qquad\qquad 6y = -4x - 2$ $\qquad\qquad 3 \cdot \frac{2}{3}x + 3 \cdot y = 3 \cdot \frac{1}{3}$

$\qquad\qquad y = \frac{-4x}{6} - \frac{2}{6}$ $\qquad\qquad 2x + 3y = 1$

$\qquad\qquad y = -\frac{2}{3}x - \frac{1}{3}$ $\qquad\qquad 3y = -2x + 1$

$\qquad m = -\frac{2}{3}, b = -\frac{1}{3}$ $\qquad\qquad y = -\frac{2}{3}x + \frac{1}{3}$

$\qquad\qquad\qquad\qquad m = -\frac{2}{3}, b = \frac{1}{3}$

The lines have the same slope but different y-intercepts, so the lines are parallel. Therefore the system is independent and inconsistent, and there is no solution.

47. $\qquad \frac{x}{2} - 2y = 1$ $\qquad\qquad y = \frac{1}{4}x - \frac{1}{2}$

$\qquad 2 \cdot \frac{x}{2} - 2 \cdot 2y = 2 \cdot 1$ $\qquad\qquad m = \frac{1}{4}, b = -\frac{1}{2}$

$\qquad\qquad x - 4y = 2$

$\qquad\qquad -4y = -x + 2$

$\qquad\qquad y = \frac{1}{4}x - \frac{1}{2}$

$m = \frac{1}{4}, b = -\frac{1}{2}$

The lines have the same slope and the same y-intercept, so the lines are the same line. Therefore the system is dependent and consistent, and there is an infinite number of solutions.

Problem Set 7.2, pp. 336-337

1. $x + y = 4$

$\underline{x - y = 2}$

$2x \quad\;\; = 6$

$\qquad x = 3$

Let $x = 3$ in $x + y = 4$.

$3 + y = 4$

$\quad\; y = 1$

The solution is $(3, 1)$.

3. $2x - y = 1$

$\underline{x + y = 5}$

$3x \quad\;\; = 6$

$\qquad x = 2$

Let $x = 2$ in $x + y = 5$.

$2 + y = 5$

$\quad\; y = 3$

The solution is $(2, 3)$.

5. $x + y = 0$

$\underline{-x + 3y = -4}$

$\qquad 4y = -4$

$\qquad\; y = -1$

Let $y = -1$ in $x + y = 0$.

$x + (-1) = 0$

$\qquad x = 1$

The solution is $(1, -1)$

7. $2x + y = 7$

$\underline{3x - y = 13}$

$5x \qquad = 20$

$\qquad x = 4$

Let $x = 4$ in $2x + y = 7$.

$2(4) + y = 7$

$8 + y = 7$

$\qquad y = -1$

The solution is $(4, -1)$.

9. $4x - 2y = -12$

$\underline{5x + 2y = -33}$

$9x \quad\quad = -45$

$x = -5$

Let $x = -5$ in $5x + 2y = -33$.

$5(-5) + 2y = -33$

$-25 + 2y = -33$

$2y = -8$

$y = -4$

The solution is $(-5, -4)$.

13. $3x - 2y = -3$

$x + y = 4$

Multiply $x + y = 4$ by 2.

$3x - 2y = -3$

$\underline{2x + 2y = 8}$

$5x \quad\quad = 5$

$x = 1$

Let $x = 1$ in $x + y = 4$.

$1 + y = 4$

$y = 3$

The solution is $(1, 3)$.

11. $-3x - 5y = -8$

$\underline{3x - 5y = 8}$

$-10y = 0$

$y = 0$

Let $y = 0$ in $3x - 5y = 8$.

$3x - 5(0) = 8$

$3x = 8$

$x = \dfrac{8}{3}$

The solution is $\left(\dfrac{8}{3}, 0\right)$.

15. $2x - y = 0$

$4x + 3y = -20$

Multiply $2x - y = 0$ by 3.

$6x - 3y = 0$

$\underline{4x + 3y = -20}$

$10x \quad\quad = -20$

$x = -2$

Let $x = -2$ in $2x - y = 0$.

$2(-2) - y = 0$

$-4 - y = 0$

$-4 = y$

The solution is $(-2, -4)$.

17. $2x + 7y = 3$

$2x + 3y = 7$

Multiply $2x + 3y = 7$ by -1.

$2x + 7y = 3$

$\underline{-2x - 3y = -7}$

$4y = -4$

$y = -1$

Let $y = -1$ in $2x + 3y = 7$.

$2x + 3(-1) = 7$

$2x - 3 = 7$

$2x = 10$

$x = 5$

The solution is $(5, -1)$.

21. $x - 2y = -5$

$4x - 3y = -20$

Multiply $x - 2y = -5$ by -4.

$-4x + 8y = 20$

$\underline{4x - 3y = -20}$

$5y = 0$

$y = 0$

Let $y = 0$ in $x - 2y = -5$.

$x - 2(0) = -5$

$x - 0 = -5$

19. $7x + 2y = 60$

$-4x + 2y = -6$

Multiply $-4x + 2y = -6$ by -1.

$7x + 2y = 60$

$\underline{4x - 2y = 6}$

$11x \qquad = 66$

$x = 6$

Let $x = 6$ in $7x + 2y = 60$.

$7(6) + 2y = 60$

$42 + 2y = 60$

$2y = 18$

$y = 9$

The solution is $(6, 9)$.

23. $4x + 9y = 6$

$2x + 3y = 1$

Multiply $2x + 3y = 1$ by -2.

$4x + 9y = 6$

$\underline{-4x - 6y = -2}$

$3y = 4$

$y = \dfrac{4}{3}$

Let $y = \dfrac{4}{3}$ in $2x + 3y = 1$.

$2x + 3(\dfrac{4}{3}) = 1$

$2x + 4 = 1$

$$x = -5$$

The solution is $(-5, 0)$.

$$2x = -3$$

$$x = -\frac{3}{2}$$

The solution is $(-\frac{3}{2}, \frac{4}{3})$.

25. $5x + 3y - 4 = 0$

$3x = 2y + 10$

Write in standard form.

$5x + 3y = 4$

$3x - 2y = 10$

Multiply $5x + 3y = 4$ by 2 and $3x - 2y = 10$ by 3.

$10x + 6y = 8$

$\underline{9x - 6y = 30}$

$19x \quad\quad = 38$

$\quad\quad x = 2$

Let $x = 2$ in $3x = 2y + 10$.

$3(2) = 2y + 10$

$6 = 2y + 10$

$-4 = 2y$

$-2 = y$

The solution is $(2, -2)$.

27. $2x + 7y - 7 = 0$

$5x + 2y - 2 = 0$

Write in standard form.

$2x + 7y = 7$

$5x + 2y = 2$

Multiply $2x + 7y = 7$ by 5 and $5x + 2y = 2$ by -2.

$10x + 35y = 35$

$\underline{-10x - 4y = -4}$

$\quad\quad 31y = 31$

$\quad\quad y = 1$

Let $y = 1$ in $5x + 2y = 2$.

$5x + 2(1) = 2$

$5x = 0$

$x = 0$

The solution is $(0, 1)$.

29. $-6x + 2y = 5$

$3x - y = 0$

Multiply $3x - y = 0$ by 2.

$-6x + 2y = 5$

$\underline{6x - 2y = 0}$

$0 = 5$

The false statement $0 = 5$
means the lines are parallel
and there is no solution.

31. $3x - 7y = 0$

$4x + 5y = 0$

Multiply $3x - 7y = 0$ by 5 and
$4x + 5y = 0$ by 7.

$15x - 35y = 0$

$\underline{28x + 35y = 0}$

$43x \qquad = 0$

$x = 0$

Let $x = 0$ in $4x + 5y = 0$.

$4(0) + 5y = 0$

$5y = 0$

$y = 0$

The solution is $(0, 0)$.

33. $x + 3y = -2$

$2x + 6y = -4$

Multiply $x + 3y = -2$ by -2.

$-2x - 6y = 4$

$\underline{2x + 6y = -4}$

$0 = 0$

The true statement $0 = 0$
means the lines are the same
and there is an infinite
number of solutions.

35. $8x + 12y = -35$

$5y = -14 - 3x$

Write in standard form.

$8x + 12y = -35$

$3x + 5y = -14$

Multiply $8x + 12y = -35$ by 3
and $3x + 5y = -14$ by -8.

$24x + 36y = -105$

$\underline{-24x - 40y = 112}$

$-4y = 7$

$y = -\dfrac{7}{4}$

Let $y = -\dfrac{7}{4}$ in $8x + 12y = -35$.

$$8x + 12\left(-\dfrac{7}{4}\right) = -35$$

$$8x - 21 = -35$$

$$8x = -14$$

$$x = -\dfrac{7}{4}$$

The solution is $\left(-\dfrac{7}{4}, -\dfrac{7}{4}\right)$.

37. $\dfrac{x}{2} + \dfrac{y}{3} = \dfrac{2}{3}$

$\dfrac{x}{3} + \dfrac{y}{5} = \dfrac{1}{3}$

Multiply the first equation by 6 and the second by 15.

$3x + 2y = 4$

$5x + 3y = 5$

Multiply $3x + 2y = 4$ by 3 and $5x + 3y = 5$ by -2.

$9x + 6y = 12$

$\underline{-10x - 6y = -10}$

$-x \qquad = 2$

$x = -2$

Let $x = -2$ in $3x + 2y = 4$.

39. $\dfrac{1}{5}x + \dfrac{1}{3}y = \dfrac{2}{5}$

$0.2x - 0.7y = 0.1$

Multiply the first equation by 15 and the second equation by 10.

$3x + 5y = 6$

$2x - 7y = 1$

Multiply $3x + 5y = 6$ by 2 and $2x - 7y = 1$ by -3.

$6x + 10y = 12$

$\underline{-6x + 21y = -3}$

$31y = 9$

$y = \dfrac{9}{31}$

Let $y = \dfrac{9}{31}$ in $2x - 7y = 1$.

$3(-2) + 2y = 4$

$-6 + 2y = 4$

$2y = 10$

$y = 5$

The solution is $(-2, 5)$.

$2x - 7\left(\dfrac{9}{31}\right) = 1$

$2x - \dfrac{63}{31} = 1$

$2x = \dfrac{94}{31}$

$x = \dfrac{47}{31}$

The solution is $\left(\dfrac{47}{31}, \dfrac{9}{31}\right)$.

41. $4.8x - 3.6y = 2.4$

$-7.2x + 5.4y = -9$

Multiply each equation by 10.

$48x - 36y = 24$

$-72x + 54y = -90$

Multiply $48x - 36y = 24$ by 54 and $-72x + 54y = -90$ by 36.

$2592x - 1944y = 1296$

$\underline{-2592x + 1944y = -3240}$

$0 = -1944$

The false statement $0 = -1944$ means the lines are parallel and there is no solution.

43. $x + y = 90$

$\underline{x - y = 38}$

$2x \qquad = 128$

$x = 64°$

Let $x = 64$ in $x + y = 90$.

$64 + y = 90$

$y = 26°$

45. $4x + y = 26$

$x - 2y = 38$

Multiply $4x + y = 26$ by 2.

$8x + 2y = 52$

$\underline{x - 2y = 38}$

$9x \qquad = 90$

$x = 10$

Let $x = 10$ in $4x + y = 26$.

$4(10) + y = 26$

$40 + y = 26$

$y = -14$

47. $\frac{1}{4}x = \frac{1}{3}y - 2$

$\frac{3}{4}y = \frac{1}{2}x + 5$

Multiply the first equation by 12 and the second by 4.

$3x = 4y - 24$

$3y = 2x + 20$

Write in standard form.

$3x - 4y = -24$

$-2x + 3y = 20$

Multiply $3x - 4y = -24$ by 2 and $-2x + 3y = 20$ by 3.

$6x - 8y = -48$

$\underline{-6x + 9y = 60}$

$y = 12$

Let $y = 12$ in $3x = 4y - 24$.

$3x = 4(12) - 24$

$3x = 24$

$x = 8$

Problem Set 7.3, pp. 342-343

1. $3x + 2y = 22$

    $y = 4x$

    Replace y with 4x in
    $3x + 2y = 22$.

    $3x + 2(4x) = 22$

    $3x + 8x = 22$

    $11x = 22$

    $x = 2$

    Let $x = 2$ in $y = 4x$.

    $y = 4(2) = 8$

    The solution is $(2, 8)$.

5. $x = 11 - 3y$

    $5x + 2y = 3$

    Replace x with $11 - 3y$ in
    $5x + 2y = 3$.

    $5(11 - 3y) + 2y = 3$

    $55 - 15y + 2y = 3$

    $-13y = -52$

    $y = 4$

    Let $y = 4$ in $x = 11 - 3y$.

    $x = 11 - 3(4) = -1$

    The solution is $(-1, 4)$.

3. $x + y = 14$

    $y = x + 2$

    Replace y with $x + 2$ in
    $x + y = 14$.

    $x + (x + 2) = 14$

    $2x + 2 = 14$

    $2x = 12$

    $x = 6$

    Let $x = 6$ in $y = x + 2$.

    $y = 6 + 2 = 8$

    The solution is $(6, 8)$.

7. $10x - 2y = 0$

    $x = \frac{1}{2}y$

    Replace x with $\frac{1}{2}y$ in
    $10x - 2y = 0$.

    $10(\frac{1}{2}y) - 2y = 0$

    $5y - 2y = 0$

    $3y = 0$

    $y = 0$

    Let $y = 0$ in $x = \frac{1}{2}y$.

    $x = \frac{1}{2}(0) = 0$

    The solution is $(0, 0)$.

9. $8x - 5y = 11$

$y + 6 = 0$

Solve $y + 6 = 0$ for y.

$y = -6$

Replace y with -6 in
$8x - 5y = 11$.

$8x - 5(-6) = 11$

$8x + 30 = 11$

$8x = -19$

$x = -\dfrac{19}{8}$

The solution is $(-\dfrac{19}{8}, -6)$.

11. $11x - 4y = -6$

$y = \dfrac{5}{4} + 3x$

Replace y with $\dfrac{5}{4} + 3x$ in
$11x - 4y = -6$.

$11x - 4(\dfrac{5}{4} + 3x) = -6$

$11x - 5 - 12x = -6$

$-x = -1$

$x = 1$

Let $x = 1$ in $y = \dfrac{5}{4} + 3x$.

$y = \dfrac{5}{4} + 3(1) = \dfrac{17}{4}$

The solution is $(1, \dfrac{17}{4})$.

13. $x - 3y = 10$

$2x + y = 6$

Solve $x - 3y = 10$ for x.

$x = 10 + 3y$

Replace x with $10 + 3y$ in
$2x + y = 6$.

$2(10 + 3y) + y = 6$

$20 + 6y + y = 6$

$7y = -14$

$y = -2$

15. $2x + y = 5$

$3x - 4y = 13$

Solve $2x + y = 5$ for y.

$y = 5 - 2x$

Replace y with $5 - 2x$ in
$3x - 4y = 13$.

$3x - 4(5 - 2x) = 13$

$3x - 20 + 8x = 13$

$11x = 33$

$x = 3$

Let $y = -2$ in $x = 10 + 3y$.

$x = 10 + 3(-2) = 4$

The solution is $(4, -2)$.

17. $8x + 4y = 7$

  $2x + y = 3$

Solve $2x + y = 3$ for y.

$y = 3 - 2x$

Replace y with $3 - 2x$ in
$8x + 4y = 7$.

$8x + 4(3 - 2x) = 7$

  $8x + 12 - 8x = 7$

      $12 = 7$

The false statement $12 = 7$ means
the lines are parallel and there
is no solution.

21. $3x + 6y = -3$

  $x + 2y = -1$

Solve $x + 2y = -1$ for x.

$x = -1 - 2y$

Replace x with $-1 - 2y$ in
$3x + 6y = -3$.

Let $x = 3$ in $y = 5 - 2x$.

$y = 5 - 2(3) = -1$

The solution is $(3, -1)$.

19. $5x - 2y = -9$

  $4x - y = -3$

Solve $4x - y = -3$ for y.

$-y = -3 - 4x$

 $y = 3 + 4x$

Replace y with $3 + 4x$ in
$5x - 2y = -9$.

$5x - 2(3 + 4x) = -9$

  $5x - 6 - 8x = -9$
         $-3x = -3$
           $x = 1$

Let $x = 1$ in $y = 3 + 4x$.

$y = 3 + 4(1) = 7$

The solution is $(1, 7)$.

23. $-3x + 5y = 6$

  $-x + 4y = 2$

Solve $-x + 4y = 2$ for x.

$-x = 2 - 4y$

 $x = -2 + 4y$

Replace x with $-2 + 4y$ in
$-3x + 5y = 6$.

$$3(-1 - 2y) + 6y = -3$$

$$-3 - 6y + 6y = -3$$

$$-3 = -3$$

The true statement $-3 = -3$ means the lines are the same line and there is an infinite number of solutions.

$$-3(-2 + 4y) + 5y = 6$$

$$6 - 12y + 5y = 6$$

$$-7y = 0$$

$$y = 0$$

Let $y = 0$ in $x = -2 + 4y$.

$$x = -2 + 4(0) = -2$$

The solution is $(-2, 0)$.

25. $4x + 3y = 7$

$$2x - 9y = 0$$

Solve $2x - 9y = 0$ for x.

$$2x = 9y$$

$$x = \frac{9}{2}y$$

Replace x with $\frac{9}{2}y$ in $4x + 3y = 7$.

$$4(\tfrac{9}{2}y) + 3y = 7$$

$$18y + 3y = 7$$

$$21y = 7$$

$$y = \frac{1}{3}$$

Let $y = \frac{1}{3}$ in $x = \frac{9}{2}y$.

27. $5x + 2y = 9$

$$3x - 4y = -5$$

Solve $5x + 2y = 9$ for y.

$$2y = 9 - 5x$$

$$y = \frac{9 - 5x}{2}$$

Replace y with $\frac{9 - 5x}{2}$ in $3x - 4y = -5$.

$$3x - 4(\frac{9 - 5x}{2}) = -5$$

$$3x - 2(9 - 5x) = -5$$

$$3x - 18 + 10x = -5$$

$$13x = 13$$

$$x = 1$$

Let $x = 1$ in $y = \frac{9 - 5x}{2}$.

$x = \frac{9}{2}(\frac{1}{3}) = \frac{3}{2}$

The solution is $(\frac{3}{2}, \frac{1}{3})$.

29. $5x + 3y = 2$

$3x - 4y = 4$

Solve $5x + 3y = 2$ for y.

$3y = 2 - 5x$

$y = \frac{2 - 5x}{3}$

Replace y with $\frac{2 - 5x}{3}$ in
$3x - 4y = 4$.

$3x - 4(\frac{2 - 5x}{3}) = 4$

Multiply by 3.

$9x - 4(2 - 5x) = 12$

$9x - 8 + 20x = 12$

$29x = 20$

$x = \frac{20}{29}$

Let $x = \frac{20}{29}$ in $y = \frac{2 - 5x}{3}$.

$y = \frac{2 - 5(\frac{20}{29})}{3}$

$y = \frac{2 - \frac{100}{29}}{3} \cdot \frac{29}{29}$

$y = \frac{9 - 5(1)}{2} = \frac{4}{2} = 2$

The solution is $(1, 2)$.

31. $\frac{x}{2} + \frac{y}{3} = 1$

$\frac{x}{4} - y = 4$

Multiply the first equation by 6
and the second by 4.

$3x + 2y = 6$

$x - 4y = 16$

Solve $x - 4y = 16$ for x.

$x = 16 + 4y$

Replace x with $16 + 4y$ in
$3x + 2y = 6$.

$3(16 + 4y) + 2y = 6$

$48 + 12y + 2y = 6$

$14y = -42$

$y = -3$

Let $y = -3$ in $x = 16 + 4y$.

$x = 16 + 4(-3) = 4$

The solution is $(4, -3)$.

$$y = \frac{58 - 100}{87}$$

$$y = -\frac{14}{29}$$

The solution is $(\frac{20}{29}, -\frac{14}{29})$.

33. $\frac{1}{3}x - \frac{1}{2}y = -\frac{11}{12}$

$\frac{1}{4}x + \frac{1}{2}y = \frac{5}{8}$

Multiply the first equation by 12 and the second by 8.

$4x - 6y = -11$

$2x + 4y = 5$

Solve the second equation for x.

$2x = 5 - 4y$

$x = \frac{5 - 4y}{2}$

Replace x with $\frac{5 - 4y}{2}$ in
$4x - 6y = -11$.

$4(\frac{5 - 4y}{2}) - 6y = -11$

$2(5 - 4y) - 6y = -11$

$10 - 8y - 6y = -11$

$-14y = -21$

$y = \frac{3}{2}$

35. $-\frac{2}{9}x + \frac{5}{18}y = \frac{1}{3}$

$\frac{1}{3}x - \frac{5}{12}y = \frac{3}{8}$

Multiply the first equation by 18 and the second by 24.

$-4x + 5y = 6$

$8x - 10y = 9$

Solve the first equation for y.

$5y = 6 + 4x$

$y = \frac{6 + 4x}{5}$

Replace y with $\frac{6 + 4x}{5}$ in
$8x - 10y = 9$.

$8x - 10(\frac{6 + 4x}{5}) = 9$

$8x - 2(6 + 4x) = 9$

$8x - 12 - 8x = 9$

$-12 = 9$

The false statement $-12 = 9$ means the lines are parallel and there is no solution.

Let $y = \frac{3}{2}$ in $x = \frac{5 - 4y}{2}$.

$$x = \frac{5 - 4(\frac{3}{2})}{2} = \frac{5 - 6}{2} = -\frac{1}{2}$$

The solution is $(-\frac{1}{2}, \frac{3}{2})$.

37. $D = 216 - 6p$

$S = 10p$

Let $S = D$.

$10p = 216 - 6p$

$16p = 216$

$p = \$13.50$

Let $p = 13.5$ in $S = 10p$.

$S = 10(13.5) = 135$ units

39. $D = 1750 - 35p$

$S = 100 + 15p$

Let $S = D$.

$100 + 15p = 1750 - 35p$

$50p = 1650$

$p = \$33$

Let $p = 33$ in $S = 100 + 15p$.

$S = 100 + 15(33) = 595$ units

41. $2x + y = 68$

$y = 6x$

Replace y with $6x$ in $2x + y = 68$.

$2x + 6x = 68$

$8x = 68$

$x = \frac{17}{2}$

Let $x = \frac{17}{2}$ in $y = 6x$.

$y = 6(\frac{17}{2}) = 51$

43. $x - y = 13$

$x = 4y - 5$

Replace x with $4y - 5$ in $x - y = 13$.

$4y - 5 - y = 13$

$3y = 18$

$y = 6$

Let $y = 6$ in $x = 4y - 5$.

$x = 4(6) - 5 = 19$

45. $x + y = 54$

$y = x + 2$

Replace y with $x + 2$ in $x + y = 54$.

$x + x + 2 = 54$

$2x = 52$

$x = 26$

Let $x = 26$ in $y = x + 2$.

$y = 26 + 2$

$y = 28$

Problem Set 7.4, pp. 348-350

1. $x = $ width

$y = $ length

$2x + 2y = 124$

$y = 4x + 7$

Solving this system gives

$x = 11$ cm and $y = 51$ cm.

5. $x + y + 27 = 180$

$x = \frac{1}{2}y$

Solving this system gives

$x = 51°$ and $y = 102°$.

3. $2x + y = 64$

$y = \frac{2}{3}x$

Solving this system gives

$x = 24$ ft and $y = 16$ ft.

7. $x = $ cost of a basketball

$y = $ cost of a football

$3x + 2y = 41$

$x + 4y = 47$

Solving this system gives

$x = \$7$ and $y = \$10$.

9. x = no. of dimes

y = no. of quarters

10x + 25y = 1045

x = y - 4

Solving this system gives

x = 27 dimes and

y = 31 quarters.

11. x = no. of nickels

y = no. of dimes

x + y = 24

5x + 10y = 175

Solving this system gives

x = 13 nickels and

y = 11 dimes.

13. x = no. of adults

y = no. of students

5x + 3y = 645

x + y = 145

Solving this system gives

x = 105 adults and

y = 40 students.

15. x = kg of 20% solution

y = kg of 60% solution

0.20x + 0.60y = 0.35(40)

x + y = 40

Solving this system gives

x = 25 kg and y = 15 kg.

17. x = lb of $0.85 nuts

y = lb of $1.35 nuts

0.85x + 1.35y = 1.00(20)

x + y = 20

Solving this system gives

x = 14 lb and y = 6 lb.

19. x = g of pure silver

y = g of 15% alloy

1.00x + 0.15y = 0.50(34)

x + y = 34

Solving this system gives

x = 14 g and y = 20 g.

21. x = amount at 5%

y = amount at 7%

$$x + y = 25{,}000$$

$$0.05x + 0.07y = 1390$$

Solving this system gives

x = \$18,000 and y = \$7000.

25. x = amount at 9½%

y = amount at 14½%

$$x = 2y$$

$$0.095x + 0.145y = 4355$$

Solving this system gives

x = \$26,000 and y = \$13,000.

29. x = distance of enemy missile

y = distance of friendly missile

$$x + y = 5500$$

$$\frac{x}{7000} = \frac{y}{8400}$$

Solving this system gives

x = 2500 mi and y = 3000 mi.

23. x = amount at 8%

y = amount at 12%

$$x + y = 6300$$

$$.08x = 0.12y$$

Solving this system gives

x = \$3780 and y = \$2520.

27. x = speed of walker

y = speed of runner

$$y = 2x$$

$$3x + 3y = 36$$

Solving this system gives

x = 4 mph and y = 8 mph.

31. c = speed of current

b = speed of boat

$$2(b - c) = 16$$

$$1\tfrac{1}{3}(b + c) = 16$$

Solving this system gives

b = 10 mph and c = 2 mph.

33. w = speed of wind

h = speed of helicopter

h + w = 102.5

h - w = 87.5

Solving this system gives

h = 95 mph and w = 7.5 mph.

Problem Set 7.5, p. 353

1. $x > 3$
   $y < 2$

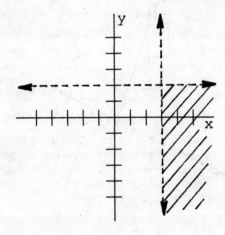

3. $x \leq -1$
   $y \leq -4$

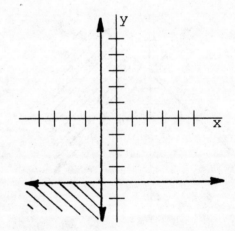

5. $x \geq 0$
   $y \geq 0$

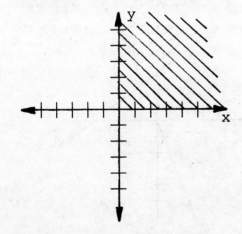

7. $y \geq x$
   $y > 2$

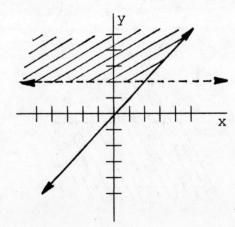

9.   x + y < 5
       x ≥ 1

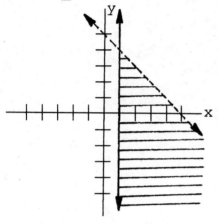

11.   5x – 3y < 10
         x ≤ 0

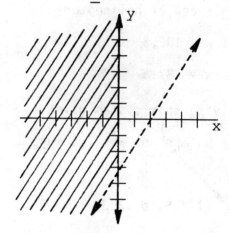

13.   x + y > 3
         y ≤ x + 3

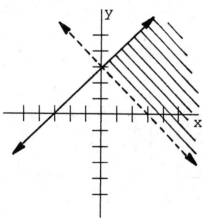

15.   3x + y ≥ 6
       x – 2y > 2

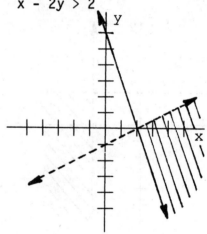

17.   x + y < 4
       2x – y ≥ -4

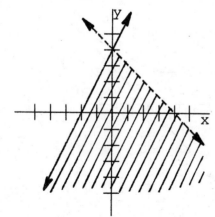

19.   y ≥ 3x
       3x + 4y ≥ 12

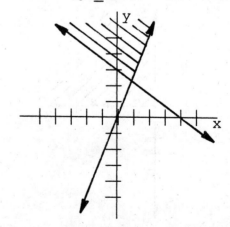

21. $y \leq x + 2$
     $y \geq x$

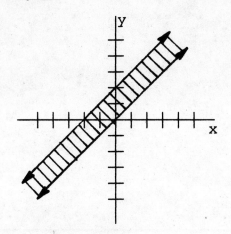

23. $y \geq x + 2$
     $y \leq x$

No solution

25. $x + 2y \leq 10$

     $x \leq \frac{1}{2}y$

     $x \geq 0$

     $y \geq 0$

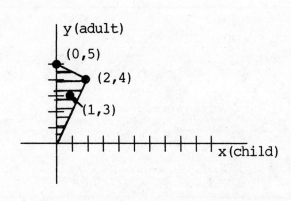

NOTES

CHAPTER 8

ROOTS AND RADICALS

Problem Set 8.1, pp. 362-366

1. The square roots of 16 are 4 and -4.

3. The square roots of 121 are 11 and -11.

5. The square roots of $\frac{1}{9}$ are $\frac{1}{3}$ and $-\frac{1}{3}$.

7. The square roots of $\frac{4}{25}$ are $\frac{2}{5}$ and $-\frac{2}{5}$.

9. The square roots of 0.49 are 0.7 and -0.7.

11. The square roots of 2809 are 53 and -53.

13. $\sqrt{4} = 2$

15. $-\sqrt{4} = -2$

17. $\pm\sqrt{4} = \pm 2$

19. $\sqrt{900} = 30$

21. $-\sqrt{1} = -1$

23. $\pm\sqrt{0} = \pm 0 = 0$

25. $\sqrt{\frac{1}{81}} = \frac{1}{9}$

27. $\sqrt{\frac{25}{144}} = \frac{5}{12}$

29. $\sqrt{-9}$ is not a real number.

31. $\sqrt{4^2} = \sqrt{16} = 4$

33. $\sqrt{(-3)^2} = \sqrt{9} = 3$

35. $-\sqrt{-81}$ is not a real number.

37. Irrational

$\sqrt{2} \approx 1.41$

39. Irrational

$\sqrt{23} \approx 4.80$

41. Rational

$\sqrt{169} = 13$

43. Irrational

$\sqrt{48} \approx 6.93$

45. Irrational

$\sqrt{177} \approx 13.30$

47. Rational

$\sqrt{361} = 19$

49. $\sqrt[3]{27} = 3$

51. $\sqrt[3]{-27} = -3$

53. $-\sqrt[3]{125} = -5$

55. $-\sqrt[3]{-125} = -(-5) = 5$

57. $\sqrt[3]{343} = 7$

59. $\sqrt[4]{16} = 2$

61. $\sqrt[4]{-16}$ is not a real number.

63. $\sqrt[5]{32} = 2$

65. $\sqrt[5]{-1} = -1$

67. $\sqrt[3]{4^3} = \sqrt[3]{64} = 4$

69. $\sqrt[6]{0} = 0$

71. $\sqrt[4]{(-3)^4} = \sqrt[4]{81} = 3$

73. $\sqrt{16 + 9} = \sqrt{25} = 5$ and $\sqrt{16} + \sqrt{9} = 4 + 3 = 7$.

Conclusion: $\sqrt{a + b} \neq \sqrt{a} + \sqrt{b}$

75. $\sqrt{625 - 49} = \sqrt{576} = 24$ and $\sqrt{625} - \sqrt{49} = 25 - 7 = 18$.

Conclusion: $\sqrt{a - b} \neq \sqrt{a} - \sqrt{b}$

77. $v = \sqrt{32r} = \sqrt{32(50)} = \sqrt{1600} = 40$ ft/sec

79. $A = s^2$

$100 = s^2$

$s = 10$ cm

81. $c = \sqrt{a^2 + b^2}$

$c = \sqrt{80^2 + 39^2}$

$c = \sqrt{6400 + 1521}$

$c = \sqrt{7921}$

$c = 89$ mi

83. $F = \sqrt{F_1^2 + F_2^2}$

$F = \sqrt{6000^2 + 8000^2}$

$F = \sqrt{36,000,000 + 64,000,000}$

$F = \sqrt{100,000,000}$

$F = 10,000$ lb

85. $s = \frac{1}{2}(a + b + c) = \frac{1}{2}(10 + 17 + 21) = \frac{1}{2}(48) = 24$

$A = \sqrt{s(s - a)(s - b)(s - c)}$

$A = \sqrt{24(24 - 10)(24 - 17)(24 - 21)}$

$A = \sqrt{24(14)(7)(3)}$

$A = \sqrt{7056}$

$A = 84$ sq in.

87. | Clear | 4913 | INV | y^x | 3 | = | 17

$\sqrt[3]{4913} = 17$

89. | Clear | 100 | INV | y^x | 3 | = | 4.64

$\sqrt[3]{100} \approx 4.64$

91. | Clear | 4096 | +/- | INV | y^x | 4 | = | Error

$\sqrt[4]{-4096}$ is not a real number.

93. | Clear | 7776 | INV | y^x | 5 | = | 6

$\sqrt[5]{7776} = 6$

Problem Set 8.2, pp. 371-373

1. $\sqrt{20} = \sqrt{4 \cdot 5} = \sqrt{4} \cdot \sqrt{5} = 2 \cdot \sqrt{5} = 2\sqrt{5}$

3. $\sqrt{18} = \sqrt{9 \cdot 2} = \sqrt{9} \cdot \sqrt{2} = 3 \cdot \sqrt{2} = 3\sqrt{2}$

5. $\sqrt{300} = \sqrt{100 \cdot 3} = \sqrt{100} \sqrt{3} = 10\sqrt{3}$

7. $\sqrt{48} = \sqrt{16 \cdot 3} = \sqrt{16} \sqrt{3} = 4\sqrt{3}$

9. $3\sqrt{50} = 3\sqrt{25}\sqrt{2} = 3 \cdot 5\sqrt{2} = 15\sqrt{2}$

11. $5\sqrt{76} = 5\sqrt{4}\sqrt{19} = 5 \cdot 2\sqrt{19} = 10\sqrt{19}$

13. $\sqrt{42}$ cannot be simplified.

15. $-4\sqrt{432} = -4\sqrt{144}\sqrt{3} = -4\cdot 12\sqrt{3} = -48\sqrt{3}$

17. $\sqrt{\dfrac{64}{121}} = \dfrac{\sqrt{64}}{\sqrt{121}} = \dfrac{8}{11}$

19. $\sqrt{\dfrac{1}{100}} = \dfrac{\sqrt{1}}{\sqrt{100}} = \dfrac{1}{10}$

21. $\sqrt{\dfrac{3}{4}} = \dfrac{\sqrt{3}}{\sqrt{4}} = \dfrac{\sqrt{3}}{2}$

23. $\sqrt{\dfrac{6}{49}} = \dfrac{\sqrt{6}}{\sqrt{49}} = \dfrac{\sqrt{6}}{7}$

25. $\sqrt{\dfrac{8}{9}} = \dfrac{\sqrt{8}}{\sqrt{9}} = \dfrac{\sqrt{4}\sqrt{2}}{3} = \dfrac{2\sqrt{2}}{3}$

27. $\dfrac{5\sqrt{63}}{3} = \dfrac{5\sqrt{9}\sqrt{7}}{3} = \dfrac{5\cdot 3\sqrt{7}}{3} = 5\sqrt{7}$

29. $\sqrt{\dfrac{245}{64}} = \dfrac{\sqrt{245}}{\sqrt{64}} = \dfrac{\sqrt{49}\sqrt{5}}{8} = \dfrac{7\sqrt{5}}{8}$

31. $-6\sqrt{\dfrac{156}{144}} = \dfrac{-6\sqrt{156}}{\sqrt{144}} = \dfrac{-6\sqrt{4}\sqrt{39}}{12} = \dfrac{-6\cdot 2\sqrt{39}}{12} = \dfrac{-12\sqrt{39}}{12} = -\sqrt{39}$

33. $\sqrt{x^4} = x^2$ Since $(x^2)^2 = x^4$

35. $\sqrt{4x^2} = \sqrt{4}\sqrt{x^2} = 2x$

37. $\sqrt{25x^2 y^6} = \sqrt{25}\sqrt{x^2}\sqrt{y^6} = 5xy^3$

39. $\sqrt{169z^{16}} = \sqrt{169}\sqrt{z^{16}} = 13z^8$

41. $\sqrt{m^7} = \sqrt{m^6\cdot m} = \sqrt{m^6}\sqrt{m} = m^3\sqrt{m}$

43. $\sqrt{r^2 s} = \sqrt{r^2}\sqrt{s} = r\sqrt{s}$

45. $\sqrt{9k^3} = \sqrt{9k^2\cdot k} = \sqrt{9k^2}\sqrt{k} = 3k\sqrt{k}$

47. $\sqrt{80p^5} = \sqrt{16p^4 \cdot 5p} = \sqrt{16p^4}\sqrt{5p} = 4p^2\sqrt{5p}$

49. $2\sqrt{500t^{13}} = 2\sqrt{100t^{12}}\sqrt{5t} = 2 \cdot 10t^6\sqrt{5t} = 20t^6\sqrt{5t}$

51. $10\sqrt{81x^6y^3} = 10\sqrt{81x^6y^2}\sqrt{y} = 10 \cdot 9x^3y\sqrt{y} = 90x^3y\sqrt{y}$

53. $-4a\sqrt{125a^{10}b^3} = -4a\sqrt{25a^{10}b^2}\sqrt{5b} = -4a \cdot 5a^5b\sqrt{5b} = -20a^6b\sqrt{5b}$

55. $\frac{2}{3}rs\sqrt{36r^3st^2} = \frac{2}{3}rs\sqrt{36r^2t^2}\sqrt{rs} = \frac{2}{3}rs \cdot 6rt\sqrt{rs} = 4r^2st\sqrt{rs}$

57. $\sqrt{\frac{y^6}{9}} = \frac{\sqrt{y^6}}{\sqrt{9}} = \frac{y^3}{3}$

59. $\sqrt{\frac{16}{x^4}} = \frac{\sqrt{16}}{\sqrt{x^4}} = \frac{4}{x^2}$

61. $\sqrt{\frac{3}{r^2}} = \frac{\sqrt{3}}{\sqrt{r^2}} = \frac{\sqrt{3}}{r}$

63. $\frac{\sqrt{24k}}{6} = \frac{\sqrt{4}\sqrt{6k}}{6} = \frac{2\sqrt{6k}}{6} = \frac{\sqrt{6k}}{3}$

65. $\sqrt{\frac{60x^6}{12}} = \sqrt{5x^6} = \sqrt{x^6}\sqrt{5} = x^3\sqrt{5}$

67. $\sqrt{\frac{27p^3}{q^2}} = \frac{\sqrt{27p^3}}{\sqrt{q^2}} = \frac{\sqrt{9p^2}\sqrt{3p}}{q} = \frac{3p\sqrt{3p}}{q}$

69. $\sqrt{\frac{5ab^2c}{20abc^3}} = \sqrt{\frac{b}{4c^2}} = \frac{\sqrt{b}}{\sqrt{4c^2}} = \frac{\sqrt{b}}{2c}$

71. $\sqrt{\frac{68x^5}{36x^3y^4}} = \sqrt{\frac{17x^2}{9y^4}} = \frac{\sqrt{17x^2}}{\sqrt{9y^4}} = \frac{\sqrt{x^2}\sqrt{17}}{\sqrt{9}\sqrt{y^4}} = \frac{x\sqrt{17}}{3y^2}$

73. $\sqrt[3]{24} = \sqrt[3]{8 \cdot 3} = \sqrt[3]{8} \cdot \sqrt[3]{3} = 2 \cdot \sqrt[3]{3} = 2\sqrt[3]{3}$

75. $\sqrt[3]{625} = \sqrt[3]{125 \cdot 5} = \sqrt[3]{125} \cdot \sqrt[3]{5} = 5 \cdot \sqrt[3]{5} = 5\sqrt[3]{5}$

77. $\sqrt[3]{250} = \sqrt[3]{125}\,\sqrt[3]{2} = 5\sqrt[3]{2}$

79. $\sqrt[4]{32} = \sqrt[4]{16}\,\sqrt[4]{2} = 2\sqrt[4]{2}$

81. $\sqrt[3]{\dfrac{9}{64}} = \dfrac{\sqrt[3]{9}}{\sqrt[3]{64}} = \dfrac{\sqrt[3]{9}}{4}$

83. $\sqrt[4]{\dfrac{80}{81}} = \dfrac{\sqrt[4]{80}}{\sqrt[4]{81}} = \dfrac{\sqrt[4]{16}\,\sqrt[4]{5}}{3} = \dfrac{2\sqrt[4]{5}}{3}$

85. $\sqrt[3]{-8x^3} = \sqrt[3]{-8}\cdot\sqrt[3]{x^3} = -2\cdot x = -2x$

87. $\sqrt[3]{64y^4} = \sqrt[3]{64y^3}\,\sqrt[3]{y} = 4y\sqrt[3]{y}$

89. $\sqrt[3]{\dfrac{9r^3s^2}{1000t^6}} = \dfrac{\sqrt[3]{9r^3s^2}}{\sqrt[3]{1000t^6}} = \dfrac{\sqrt[3]{r^3}\,\sqrt[3]{9s^2}}{10t^2} = \dfrac{r\sqrt[3]{9s^2}}{10t^2}$

91. $\sqrt[3]{\dfrac{16m^9n^2}{27mn^5}} = \sqrt[3]{\dfrac{16m^8}{27n^3}} = \dfrac{\sqrt[3]{16m^8}}{\sqrt[3]{27n^3}} = \dfrac{\sqrt[3]{8m^6}\,\sqrt[3]{2m^2}}{3n} = \dfrac{2m^2\sqrt[3]{2m^2}}{3n}$

93. $\sqrt{(x + 3)^2} = x + 3$

95. $\sqrt{x^2 + 4x + 4} = \sqrt{(x + 2)^2} = x + 2$

97. $\sqrt{9x^2 + 30xy + 25y^2} = \sqrt{(3x + 5y)^2} = 3x + 5y$

99. $s_a = s_t\sqrt{\dfrac{\ell_a}{\ell_t}}$

$s_a = 40\sqrt{\dfrac{90}{40}} = 40\sqrt{\dfrac{9}{4}} = 40\left(\dfrac{3}{2}\right) = 60 \text{ mph}$

101. $c = \sqrt{a^2 + b^2}$

$c = \sqrt{65^2 + 65^2} = \sqrt{65^2\cdot 2} = \sqrt{65^2}\cdot\sqrt{2} = 65\sqrt{2} \text{ ft}$

103. $A = s^2$

$160 = s^2$

$s = \sqrt{160} = \sqrt{16}\,\sqrt{10} = 4\sqrt{10} \text{ m}$

Problem Set 8.3, pp. 376-378

1. $\sqrt{2}\sqrt{3} = \sqrt{2 \cdot 3} = \sqrt{6}$

3. $\sqrt{6}\sqrt{7} = \sqrt{6 \cdot 7} = \sqrt{42}$

5. $\sqrt{14}\sqrt{m} = \sqrt{14m}$

7. $\sqrt{5}\sqrt{10} = \sqrt{5 \cdot 10} = \sqrt{50} = \sqrt{25}\sqrt{2} = 5\sqrt{2}$

9. $\sqrt{6}\sqrt{10} = \sqrt{6 \cdot 10} = \sqrt{60} = \sqrt{4}\sqrt{15} = 2\sqrt{15}$

11. $\sqrt{12}\sqrt{15} = \sqrt{12 \cdot 15} = \sqrt{180} = \sqrt{36}\sqrt{5} = 6\sqrt{5}$

13. $\sqrt{2}\sqrt{2} = 2$

15. $(\sqrt{7})^2 = 7$

17. $\sqrt{x}\sqrt{x} = x$

19. $(3\sqrt{2})(5\sqrt{11}) = (3 \cdot 5)(\sqrt{2}\sqrt{11}) = 15\sqrt{22}$

21. $(2\sqrt{3})(4\sqrt{27}) = (2 \cdot 4)(\sqrt{3}\sqrt{27}) = 8\sqrt{81} = 8 \cdot 9 = 72$

23. $\sqrt{4}\sqrt{18} = \sqrt{72} = \sqrt{36}\sqrt{2} = 6\sqrt{2}$

25. $\sqrt{x}\sqrt{x^3} = \sqrt{x \cdot x^3} = \sqrt{x^4} = x^2$

27. $\sqrt{3z}\sqrt{5z} = \sqrt{3z \cdot 5z} = \sqrt{15z^2} = \sqrt{z^2}\sqrt{15} = z\sqrt{15}$

29. $(-2\sqrt{p})^2 = (-2)^2(\sqrt{p})^2 = 4p$

31. $(6\sqrt{m})(-3\sqrt{m}) = 6(-3)\sqrt{m}\sqrt{m} = -18m$

33. $(5\sqrt{2t})^2 = 5^2(\sqrt{2t})^2 = 25(2t) = 50t$

35. $(4\sqrt{7x})(5\sqrt{11x}) = 4 \cdot 5\sqrt{7x}\sqrt{11x} = 20\sqrt{77x^2} = 20\sqrt{x^2}\sqrt{77} = 20x\sqrt{77}$

37. $\sqrt{8x^2y^2}\sqrt{2x^3y} = \sqrt{16x^5y^3} = \sqrt{16x^4y^2}\sqrt{xy} = 4x^2y\sqrt{xy}$

39. $-\sqrt{12x^5}\sqrt{4x^5} = -\sqrt{48x^{10}} = -\sqrt{16x^{10}}\sqrt{3} = -4x^5\sqrt{3}$

41. $2\sqrt{15x^{11}y^7}\sqrt{10x^5y^7} = 2\sqrt{150x^{16}y^{14}} = 2\sqrt{25x^{16}y^{14}}\sqrt{6}$

$$= 2 \cdot 5x^8y^7\sqrt{6}$$

$$= 10x^8y^7\sqrt{6}$$

43. $\sqrt{3}\sqrt{2x + 5} = \sqrt{3(2x + 5)} = \sqrt{6x + 15}$

45. $(\sqrt{y + 1})^2 = y + 1$

47. $(2\sqrt{4p + 9})^2 = 2^2(\sqrt{4p + 9})^2 = 4(4p + 9) = 16 + 36$

49. $\dfrac{\sqrt{20}}{\sqrt{5}} = \sqrt{\dfrac{20}{5}} = \sqrt{4} = 2$

51. $\dfrac{\sqrt{288}}{\sqrt{2}} = \sqrt{\dfrac{288}{2}} = \sqrt{144} = 12$

53. $\dfrac{\sqrt{15}}{\sqrt{3}} = \sqrt{\dfrac{15}{3}} = \sqrt{5}$

55. $\dfrac{\sqrt{21m}}{\sqrt{7}} = \sqrt{\dfrac{21m}{7}} = \sqrt{3m}$

57. $\dfrac{\sqrt{98r}}{\sqrt{2r}} = \sqrt{\dfrac{98r}{2r}} = \sqrt{49} = 7$

59. $\dfrac{\sqrt{75a}}{\sqrt{3}} = \sqrt{\dfrac{75a}{3}} = \sqrt{25a} = \sqrt{25}\sqrt{a} = 5\sqrt{a}$

61. $\dfrac{\sqrt{120}}{\sqrt{10}} = \sqrt{\dfrac{120}{10}} = \sqrt{12} = \sqrt{4}\sqrt{3} = 2\sqrt{3}$

63. $\dfrac{\sqrt{162k}}{\sqrt{6}} = \sqrt{\dfrac{162k}{6}} = \sqrt{27k} = \sqrt{9}\sqrt{3k} = 3\sqrt{3k}$

65. $\dfrac{12\sqrt{17}}{\sqrt{16}} = \dfrac{12\sqrt{17}}{4} = 3\sqrt{17}$

67. $\dfrac{6\sqrt{4}}{2\sqrt{2}} = \dfrac{6}{2} \cdot \dfrac{\sqrt{4}}{\sqrt{2}} = 3 \cdot \sqrt{\dfrac{4}{2}} = 3\sqrt{2}$

69. $\dfrac{8\sqrt{45x^4}}{4\sqrt{5x}} = \dfrac{8}{4} \cdot \dfrac{\sqrt{45x^4}}{\sqrt{5x}} = 2\sqrt{\dfrac{45x^4}{5x}} = 2\sqrt{9x^3}$

$$= 2\sqrt{9x^2}\sqrt{x}$$

$$= 2 \cdot 3x\sqrt{x}$$

$$= 6x\sqrt{x}$$

71. $\dfrac{9\sqrt{50y^5z^6}}{18\sqrt{2yz}} = \dfrac{9}{18} \cdot \dfrac{\sqrt{50y^5z^6}}{\sqrt{2yz}} = \dfrac{1}{2} \cdot \sqrt{\dfrac{50y^5z^6}{2yz}} = \dfrac{1}{2} \cdot \sqrt{25y^4z^5}$

$$= \dfrac{1}{2} \cdot \sqrt{25y^4z^4}\sqrt{z}$$

$$= \dfrac{1}{2} \cdot 5y^2z^2\sqrt{z}$$

$$= \dfrac{5}{2}y^2z^2\sqrt{z}$$

73. $\dfrac{\sqrt{21x^5y}}{\sqrt{28xy}} = \sqrt{\dfrac{21x^5y}{28xy}} = \sqrt{\dfrac{3x^4}{4}} = \dfrac{\sqrt{3x^4}}{\sqrt{4}} = \dfrac{\sqrt{x^4}\sqrt{3}}{2} = \dfrac{x^2\sqrt{3}}{2}$

75. $\dfrac{\sqrt{12x^2}\sqrt{18xy}}{\sqrt{6x^3y^3}} = \dfrac{\sqrt{216x^3y}}{\sqrt{6x^3y^3}} = \sqrt{\dfrac{216x^3y}{6x^3y^3}} = \sqrt{\dfrac{36}{y^2}} = \dfrac{\sqrt{36}}{\sqrt{y^2}} = \dfrac{6}{y}$

77. $\sqrt[3]{9}\sqrt[3]{3} = \sqrt[3]{9 \cdot 3} = \sqrt[3]{27} = 3$

79. $\sqrt[3]{4x^2}\sqrt[3]{16x} = \sqrt[3]{4x^2 \cdot 16x} = \sqrt[3]{64x^3} = 4x$

81. $\sqrt[3]{12y^4}\sqrt[3]{2y^2} = \sqrt[3]{24y^6} = \sqrt[3]{8y^6}\sqrt[3]{3} = 2y^2\sqrt[3]{3}$

83. $\dfrac{\sqrt[3]{250}}{\sqrt[3]{2}} = \sqrt[3]{\dfrac{250}{2}} = \sqrt[3]{125} = 5$

85. $\dfrac{\sqrt[4]{48}}{\sqrt[4]{3}} = \sqrt[4]{\dfrac{48}{3}} = \sqrt[4]{16} = 2$

87. $\dfrac{\sqrt[4]{2500x^{10}}}{\sqrt[4]{4x^2}} = \sqrt[4]{\dfrac{2500x^{10}}{4x^2}} = \sqrt[4]{625x^8} = 5x^2$

89. $a = \sqrt{c^2 - b^2}$

$a = \sqrt{(\sqrt{40})^2 - 6^2} = \sqrt{40 - 36} = \sqrt{4} = 2$

91. $a = \sqrt{c^2 - b^2}$

$a = \sqrt{5^2 - (\sqrt{7})^2} = \sqrt{25 - 7} = \sqrt{18} = \sqrt{9}\sqrt{2} = 3\sqrt{2}$

93. P = 4s

P = $4(5\sqrt{7}) = (4 \cdot 5)\sqrt{7} = 20\sqrt{7}$ yd

95. $\overline{AC}^2 = 8^2 + 24^2$

$\overline{AC}^2 = 64 + 576$

$\overline{AC}^2 = 640$

$\overline{AD}^2 = \overline{AC}^2 + 6^2$

$\overline{AD}^2 = 640 + 36$

$\overline{AD}^2 = 676$

$\overline{AD} = \sqrt{676} = 26$ in.

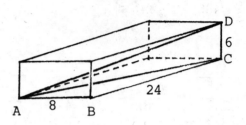

Problem Set 8.4, pp. 380-381

1. $3\sqrt{5} + 4\sqrt{5} = (3 + 4)\sqrt{5} = 7\sqrt{5}$

3. $2\sqrt{11} + \sqrt{11} = 2\sqrt{11} + 1\sqrt{11} = (2 + 1)\sqrt{11} = 3\sqrt{11}$

5. $\sqrt{3} + \sqrt{7}$ cannot be simplified.

7. $9\sqrt{15} - 4\sqrt{15} = (9 - 4)\sqrt{15} = 5\sqrt{15}$

9. $\sqrt{2} - 5\sqrt{2} = 1\sqrt{2} - 5\sqrt{2} = (1 - 5)\sqrt{2} = -4\sqrt{2}$

11. $4\sqrt{5} - 10\sqrt{5} + \sqrt{5} = (4 - 10 + 1)\sqrt{5} = -5\sqrt{5}$

13. $3\sqrt{2} - 4\sqrt{5} + 2\sqrt{2} = (3\sqrt{2} + 2\sqrt{2}) - 4\sqrt{5} = 5\sqrt{2} - 4\sqrt{5}$

15. $4.2\sqrt{17} + 3.6\sqrt{17} = (4.2 + 3.6)\sqrt{17} = 7.8\sqrt{17}$

17. $-\frac{2}{3}\sqrt{15} - \frac{3}{4}\sqrt{15} = (-\frac{2}{3} - \frac{3}{4})\sqrt{15} = (-\frac{8}{12} - \frac{9}{12})\sqrt{15} = -\frac{17}{12}\sqrt{15}$

19. $\sqrt{20} + \sqrt{5} = \sqrt{4}\sqrt{5} + \sqrt{5} = 2\sqrt{5} + 1\sqrt{5} = 3\sqrt{5}$

21. $\sqrt{8} + \sqrt{50} = \sqrt{4}\sqrt{2} + \sqrt{25}\sqrt{2} = 2\sqrt{2} + 5\sqrt{2} = 7\sqrt{2}$

23. $\sqrt{18} + 7\sqrt{2} = \sqrt{9}\sqrt{2} + 7\sqrt{2} = 3\sqrt{2} + 7\sqrt{2} = 10\sqrt{2}$

25. $2\sqrt{63} - \sqrt{700} = 2\sqrt{9}\sqrt{7} - \sqrt{100}\sqrt{7} = 2 \cdot 3\sqrt{7} - 10\sqrt{7}$

$$= 6\sqrt{7} - 10\sqrt{7}$$

$$= -4\sqrt{7}$$

27. $\frac{2}{3}\sqrt{486} + \sqrt{294} = \frac{2}{3}\sqrt{81}\sqrt{6} + \sqrt{49}\sqrt{6} = \frac{2}{3} \cdot 9\sqrt{6} + 7\sqrt{6}$

$$= 6\sqrt{6} + 7\sqrt{6}$$

$$= 13\sqrt{6}$$

29. $\sqrt{24} + \sqrt{6} - \sqrt{54} = \sqrt{4}\sqrt{6} + \sqrt{6} - \sqrt{9}\sqrt{6} = 2\sqrt{6} + \sqrt{6} - 3\sqrt{6} = 0\sqrt{6} = 0$

31. $3\sqrt{32} + \sqrt{8} - 5\sqrt{3} = 3\sqrt{16}\sqrt{2} + \sqrt{4}\sqrt{2} - 5\sqrt{3} = 3 \cdot 4\sqrt{2} + 2\sqrt{2} - 5\sqrt{3}$

$$= 12\sqrt{2} + 2\sqrt{2} - 5\sqrt{3}$$

$$= 14\sqrt{2} - 5\sqrt{3}$$

33. $2\sqrt{2}\sqrt{5} - 3\sqrt{10} + \sqrt{40} = 2\sqrt{10} - 3\sqrt{10} + \sqrt{4}\sqrt{10} = 2\sqrt{10} - 3\sqrt{10} + 2\sqrt{10}$

$$= 1\sqrt{10}$$

$$= \sqrt{10}$$

35. $\sqrt{3}\sqrt{15} + \sqrt{50} - 4\sqrt{2}\sqrt{10} = \sqrt{45} + \sqrt{50} - 4\sqrt{20} = \sqrt{9}\sqrt{5} + \sqrt{25}\sqrt{2} - 4\sqrt{4}\sqrt{5}$

$$= 3\sqrt{5} + 5\sqrt{2} - 4 \cdot 2\sqrt{5}$$

$$= 3\sqrt{5} + 5\sqrt{2} - 8\sqrt{5}$$

$$= 5\sqrt{2} - 5\sqrt{5}$$

37. $\sqrt{3a} + \sqrt{3a} = 1\sqrt{3a} + 1\sqrt{3a} = (1 + 1)\sqrt{3a} = 2\sqrt{3a}$

39. $2\sqrt{6r} - 5\sqrt{6r} + 7\sqrt{6r} = (2 - 5 + 7)\sqrt{6r} = 4\sqrt{6r}$

41. $6\sqrt{x} + x - \sqrt{x} + 2x = (6\sqrt{x} - \sqrt{x}) + (x + 2x) = 5\sqrt{x} + 3x$

43. $3z\sqrt{z} + 5z\sqrt{z} = (3z + 5z)\sqrt{z} = 8z\sqrt{z}$

45. $\sqrt{12m} + \sqrt{3m} = \sqrt{4}\sqrt{3m} + \sqrt{3m} = 2\sqrt{3m} + 1\sqrt{3m} = 3\sqrt{3m}$

47. $5\sqrt{32y} - 4\sqrt{18y} = 5\sqrt{16}\sqrt{2y} - 4\sqrt{9}\sqrt{2y} = 5\cdot4\sqrt{2y} - 4\cdot3\sqrt{2y}$

$$= 20\sqrt{2y} - 12\sqrt{2y}$$

$$= 8\sqrt{2y}$$

49. $p\sqrt{28} + \sqrt{7p^2} = p\sqrt{4}\sqrt{7} + \sqrt{p^2}\sqrt{7} = 2p\sqrt{7} + p\sqrt{7} = (2p + p)\sqrt{7}$

$$= 3p\sqrt{7}$$

51. $3\sqrt{8x^2y} - 2\sqrt{2x^2y} = 3\sqrt{4x^2}\sqrt{2y} - 2\sqrt{x^2}\sqrt{2y} = 3\cdot2x\sqrt{2y} - 2\cdot x\sqrt{2y}$

$$= 6x\sqrt{2y} - 2x\sqrt{2y}$$

$$= (6x - 2x)\sqrt{2y}$$

$$= 4x\sqrt{2y}$$

53. $\sqrt{98k^3} - \sqrt{288k^3} = \sqrt{49k^2}\sqrt{2k} - \sqrt{144k^2}\sqrt{2k} = 7k\sqrt{2k} - 12k\sqrt{2k}$

$$= (7k - 12k)\sqrt{2k}$$

$$= -5k\sqrt{2k}$$

55. $\sqrt{22x} + \sqrt{12y} - \sqrt{198x} + \sqrt{108y} = \sqrt{22x} + \sqrt{4}\sqrt{3y} - \sqrt{9}\sqrt{22x} + \sqrt{36}\sqrt{3y}$

$$= \sqrt{22x} + 2\sqrt{3y} - 3\sqrt{22x} + 6\sqrt{3y}$$

$$= -2\sqrt{22x} + 8\sqrt{3y}$$

57. $4a\sqrt{5b} + \sqrt{5a^2b} - \sqrt{20a^2b} = 4a\sqrt{5b} + \sqrt{a^2}\sqrt{5b} - \sqrt{4a^2}\sqrt{5b}$

$$= 4a\sqrt{5b} + a\sqrt{5b} - 2a\sqrt{5b}$$

$$= (4a + a - 2a)\sqrt{5b}$$

$$= 3a\sqrt{5b}$$

59. $2\sqrt{ab^3} + 3a\sqrt{ab} - 5b\sqrt{4ab} = 2\sqrt{b^2}\sqrt{ab} + 3a\sqrt{ab} - 5b\sqrt{4}\sqrt{ab}$

$$= 2b\sqrt{ab} + 3a\sqrt{ab} - 5b\cdot2\sqrt{ab}$$

$$= 2b\sqrt{ab} + 3a\sqrt{ab} - 10b\sqrt{ab}$$

$$= 3a\sqrt{ab} - 8b\sqrt{ab}$$

61. $11\sqrt[3]{2} + 2\sqrt[3]{2} + 3\sqrt{2} = (11 + 2)\sqrt[3]{2} + 3\sqrt{2} = 13\sqrt[3]{2} + 3\sqrt{2}$

63. $\sqrt[4]{7} - \sqrt[3]{7} + 6\sqrt[4]{7} = (1\sqrt[4]{7} + 6\sqrt[4]{7}) - \sqrt[3]{7} = 7\sqrt[4]{7} - \sqrt[3]{7}$

65. $\sqrt[3]{81} + 4\sqrt[3]{3} = \sqrt[3]{27}\sqrt[3]{3} + 4\sqrt[3]{3} = 3\sqrt[3]{3} + 4\sqrt[3]{3} = 7\sqrt[3]{3}$

67. $7\sqrt[3]{40x^3} - \sqrt[3]{135x^3} + \sqrt[3]{5x^3} = 7\sqrt[3]{8x^3}\sqrt[3]{5} - \sqrt[3]{27x^3}\sqrt[3]{5} + \sqrt[3]{x^3}\sqrt[3]{5}$

$$= 7\cdot2x\sqrt[3]{5} - 3x\sqrt[3]{5} + x\sqrt[3]{5}$$

$$= 14x\sqrt[3]{5} - 3x\sqrt[3]{5} + x\sqrt[3]{5}$$

$$= (14x - 3x + x)\sqrt[3]{5}$$

$$= 12x\sqrt[3]{5}$$

69. $\sqrt[4]{32y^3} + \sqrt[4]{162y^3} = \sqrt[4]{16}\sqrt[4]{2y^3} + \sqrt[4]{81}\sqrt[4]{2y^3} = 2\sqrt[4]{2y^3} + 3\sqrt[4]{2y^3} = 5\sqrt[4]{2y^3}$

71. $5\sqrt[3]{4a^4b^8} - a\sqrt[3]{4ab^8} = 5\sqrt[3]{a^3b^6}\sqrt[3]{4ab^2} - a\sqrt[3]{b^6}\sqrt[3]{4ab^2}$

$$= 5\cdot ab^2\sqrt[3]{4ab^2} - a\cdot b^2\sqrt[3]{4ab^2}$$

$$= (5ab^2 - ab^2)\sqrt[3]{4ab^2}$$

$$= 4ab^2\sqrt[3]{4ab^2}$$

73. $P = 2\ell + 2w$

$$P = 2\sqrt{242} + 2\sqrt{128} = 2\sqrt{121}\sqrt{2} + 2\sqrt{64}\sqrt{2} = 2\cdot11\sqrt{2} + 2\cdot8\sqrt{2}$$

$$= 22\sqrt{2} + 16\sqrt{2}$$

$$= 38\sqrt{2}$$

75. The first appliance requires a voltage of

$$V = \sqrt{WR} = \sqrt{300\cdot18} = \sqrt{5400} = \sqrt{900}\sqrt{6} = 30\sqrt{6} \text{ volts.}$$

The second appliance requires a voltage of

$$V = \sqrt{WR} = \sqrt{196\cdot6} = \sqrt{196}\sqrt{6} = 14\sqrt{6} \text{ volts.}$$

The total voltage is

$$30\sqrt{6} + 14\sqrt{6} = 44\sqrt{6} \text{ volts.}$$

Problem Set 8.5, pp. 383-384

1. $2(4\sqrt{5} - 1) = 2\cdot4\sqrt{5} - 2\cdot1 = 8\sqrt{5} - 2$

3. $\sqrt{3}(\sqrt{2} + \sqrt{3}) = \sqrt{3}\cdot\sqrt{2} + \sqrt{3}\cdot\sqrt{3} = \sqrt{6} + 3$

5. $4\sqrt{5}(\sqrt{3} + 2\sqrt{7}) = 4\sqrt{5}\cdot\sqrt{3} + 4\sqrt{5}\cdot2\sqrt{7} = 4\sqrt{15} + 8\sqrt{35}$

7. $5\sqrt{2}(\sqrt{8} - 2\sqrt{6}) = 5\sqrt{2}\cdot\sqrt{8} - 5\sqrt{2}\cdot2\sqrt{6} = 5\sqrt{16} - 10\sqrt{12}$

$$= 5\cdot4 - 10\sqrt{4}\sqrt{3}$$

$$= 20 - 10\cdot2\sqrt{3}$$

$$= 20 - 20\sqrt{3}$$

9. $\sqrt{x}(\sqrt{x} + \sqrt{y}) = \sqrt{x}\sqrt{x} + \sqrt{x}\sqrt{y} = x + \sqrt{xy}$

11. $\sqrt{3p}(\sqrt{27p} - \sqrt{15p}) = \sqrt{3p}\sqrt{27p} - \sqrt{3p}\sqrt{15p} = \sqrt{81p^2} - \sqrt{45p^2}$

$$= 9p - \sqrt{9p^2}\sqrt{5}$$

$$= 9p - 3p\sqrt{5}$$

13. $6\sqrt{2m}(\sqrt{14} + 7\sqrt{m}) = 6\sqrt{2m}\sqrt{14} + 6\sqrt{2m}\cdot7\sqrt{m} = 6\sqrt{28m} + 42\sqrt{2m^2}$

$$= 6\sqrt{4}\sqrt{7m} + 42\sqrt{m^2}\sqrt{2}$$

$$= 6\cdot2\sqrt{7m} + 42\cdot m\sqrt{2}$$

$$= 12\sqrt{7m} + 42m\sqrt{2}$$

15. $\sqrt{x + 1}(\sqrt{x + 1} - \sqrt{x}) = \sqrt{x + 1}\sqrt{x + 1} - \sqrt{x + 1}\sqrt{x} = x + 1 - \sqrt{(x + 1)x}$

$$= x + 1 - \sqrt{x^2 + x}$$

17. $9(2\sqrt{x} + \sqrt{y} - 4) = 9\cdot2\sqrt{x} + 9\sqrt{y} - 9\cdot4 = 18\sqrt{x} + 9\sqrt{y} - 36$

19. $(\sqrt{2} + 1)(\sqrt{2} + 3) = \sqrt{2}\sqrt{2} + 3\sqrt{2} + 1\cdot\sqrt{2} + 1\cdot3 = 2 + 3\sqrt{2} + \sqrt{2} + 3$

$$= 5 + 4\sqrt{2}$$

21. $(\sqrt{3} + 2)(\sqrt{5} - 4) = \sqrt{3}\sqrt{5} - 4\sqrt{3} + 2\sqrt{5} - 2\cdot4 = \sqrt{15} - 4\sqrt{3} + 2\sqrt{5} - 8$

23. $(\sqrt{5} + \sqrt{2})(\sqrt{3} + \sqrt{2}) = \sqrt{5}\sqrt{3} + \sqrt{5}\sqrt{2} + \sqrt{2}\sqrt{3} + \sqrt{2}\sqrt{2} = \sqrt{15} + \sqrt{10} + \sqrt{6} + 2$

25. $(\sqrt{6} + 5\sqrt{2})(\sqrt{3} - \sqrt{2}) = \sqrt{6}\sqrt{3} - \sqrt{6}\sqrt{2} + 5\sqrt{2}\sqrt{3} - 5\sqrt{2}\sqrt{2}$

$$= \sqrt{18} - \sqrt{12} + 5\sqrt{6} - 5 \cdot 2$$

$$= \sqrt{9}\sqrt{2} - \sqrt{4}\sqrt{3} + 5\sqrt{6} - 10$$

$$= 3\sqrt{2} - 2\sqrt{3} + 5\sqrt{6} - 10$$

27. $(\sqrt{14} - 1)^2 = (\sqrt{14})^2 - 2\sqrt{14 \cdot 1} + 1^2 = 14 - 2\sqrt{14} + 1 = 15 - 2\sqrt{14}$

29. $(2\sqrt{10} + 3)^2 = (2\sqrt{10})^2 + 2 \cdot 2\sqrt{10} \cdot 3 + 3^2 = 4 \cdot 10 + 12\sqrt{10} + 9 = 49 + 12\sqrt{10}$

31. $(\sqrt{7} - 3)(\sqrt{7} + 3) = (\sqrt{7})^2 - 3^2 = 7 - 9 = -2$

33. $(\sqrt{11} + \sqrt{5})(\sqrt{11} - \sqrt{5}) = (\sqrt{11})^2 - (\sqrt{5})^2 = 11 - 5 = 6$

35. $(\sqrt{2x} + 4)^2 = (\sqrt{2x})^2 + 2\sqrt{2x} \cdot 4 + 4^2 = 2x + 8\sqrt{2x} + 16$

37. $(2\sqrt{x} + 5)(3\sqrt{x} - 8) = 2\sqrt{x} \cdot 3\sqrt{x} - 2\sqrt{x} \cdot 8 + 5 \cdot 3\sqrt{x} - 5 \cdot 8$

$$= 6x - 16\sqrt{x} + 15\sqrt{x} - 40$$

$$= 6x - \sqrt{x} - 40$$

39. $(4\sqrt{m} - \sqrt{n})^2 = (4\sqrt{m})^2 - 2 \cdot 4\sqrt{m}\sqrt{n} + (\sqrt{n})^2 = 16m - 8\sqrt{mn} + n$

41. $(z - 2\sqrt{7})(z + 2\sqrt{7}) = z^2 - (2\sqrt{7})^2 = z^2 - 4 \cdot 7 = z^2 - 28$

43. $(\sqrt{3x} + \sqrt{8y})(\sqrt{2x} - \sqrt{2y}) = \sqrt{3x}\sqrt{2x} - \sqrt{3x}\sqrt{2y} + \sqrt{8y}\sqrt{2x} - \sqrt{8y}\sqrt{2y}$

$$= \sqrt{6x^2} - \sqrt{6xy} + \sqrt{16xy} - \sqrt{16y^2}$$

$$= \sqrt{x^2}\sqrt{6} - \sqrt{6xy} + \sqrt{16}\sqrt{xy} - 4y$$

$$= x\sqrt{6} - \sqrt{6xy} + 4\sqrt{xy} - 4y$$

45. $(4\sqrt{r} + 3\sqrt{6})^2 = (4\sqrt{r})^2 + 2 \cdot 4\sqrt{r} \cdot 3\sqrt{6} + (3\sqrt{6})^2 = 16r + 24\sqrt{6r} + 9 \cdot 6$

$$= 16r + 24\sqrt{6r} + 54$$

47. $(\sqrt{20k} + \sqrt{10})(\sqrt{2k} + \sqrt{3}) = \sqrt{20k}\sqrt{2k} + \sqrt{20k}\sqrt{3} + \sqrt{10}\sqrt{2k} + \sqrt{10}\sqrt{3}$

$$= \sqrt{40k^2} + \sqrt{60k} + \sqrt{20k} + \sqrt{30}$$

$$= \sqrt{4k^2}\sqrt{10} + \sqrt{4}\sqrt{15k} + \sqrt{4}\sqrt{5k} + \sqrt{30}$$

$$= 2k\sqrt{10} + 2\sqrt{15k} + 2\sqrt{5k} + \sqrt{30}$$

49. $(4\sqrt{t} - \sqrt{8})(3\sqrt{t} + \sqrt{2}) = 4\sqrt{t}\cdot 3\sqrt{t} + 4\sqrt{t}\sqrt{2} - \sqrt{8}\cdot 3\sqrt{t} - \sqrt{8}\sqrt{2}$

$$= 12\sqrt{t^2} + 4\sqrt{2t} - 3\sqrt{8t} - \sqrt{16}$$

$$= 12t + 4\sqrt{2t} - 3\sqrt{4}\sqrt{2t} - 4$$

$$= 12t + 4\sqrt{2t} - 3\cdot 2\sqrt{2t} - 4$$

$$= 12t + 4\sqrt{2t} - 6\sqrt{2t} - 4$$

$$= 12t - 2\sqrt{2t} - 4$$

51. $(\sqrt{x} - \sqrt{x + 3})^2 = (\sqrt{x})^2 - 2\sqrt{x}\sqrt{x + 3} + (\sqrt{x + 3})^2 = x - 2\sqrt{x(x + 3)} + x + 3$

$$= 2x - 2\sqrt{x^2 + 3x} + 3$$

53. $(\sqrt{y + 8} + 6)^2 = (\sqrt{y + 8})^2 + 2\sqrt{y + 8}\cdot 6 + 6^2 = y + 8 + 12\sqrt{y + 8} + 36$

$$= y + 44 + 12\sqrt{y + 8}$$

55. $\dfrac{5 - 5\sqrt{7}}{5} = \dfrac{5(1 - \sqrt{7})}{5} = 1 - \sqrt{7}$

57. $\dfrac{2 + 6\sqrt{5}}{2} = \dfrac{2(1 + 3\sqrt{5})}{2} = 1 + 3\sqrt{5}$

59. $\dfrac{3 - \sqrt{45}}{18} = \dfrac{3 - \sqrt{9}\sqrt{5}}{18} = \dfrac{3 - 3\sqrt{5}}{18} = \dfrac{3(1 - \sqrt{5})}{18} = \dfrac{1 - \sqrt{5}}{6}$

61. $\dfrac{10 + 2\sqrt{12}}{6} = \dfrac{10 + 2\sqrt{4}\sqrt{3}}{6} = \dfrac{10 + 4\sqrt{3}}{6} = \dfrac{2(5 + 2\sqrt{3})}{6} = \dfrac{5 + 2\sqrt{3}}{3}$

63. $\dfrac{8 \pm \sqrt{176}}{12} = \dfrac{8 \pm \sqrt{16}\sqrt{11}}{12} = \dfrac{8 \pm 4\sqrt{11}}{12} = \dfrac{4(2 \pm \sqrt{11})}{12} = \dfrac{2 \pm \sqrt{11}}{3}$

65. $\dfrac{-14 \pm \sqrt{245}}{28} = \dfrac{-14 \pm \sqrt{49}\sqrt{5}}{28} = \dfrac{-14 \pm 7\sqrt{5}}{28} = \dfrac{7(-2 \pm \sqrt{5})}{28} = \dfrac{-2 \pm \sqrt{5}}{4}$

67. $\sqrt[3]{2}(\sqrt[3]{4} - 5) = \sqrt[3]{2}\cdot\sqrt[3]{4} - \sqrt[3]{2}\cdot 5 = \sqrt[3]{8} - 5\sqrt[3]{2} = 2 - 5\sqrt[3]{2}$

69. $3\sqrt[4]{9}(\sqrt[4]{9} + \sqrt[4]{3}) = 3\sqrt[4]{9}\,\sqrt[4]{9} + 3\sqrt[4]{9}\,\sqrt[4]{3} = 3\sqrt[4]{81} + 3\sqrt[4]{27}$

$$= 3\cdot 3 + 3\sqrt[4]{27}$$
$$= 9 + 3\sqrt[4]{27}$$

71. $(\sqrt[3]{5} + 7)(\sqrt[3]{25} - 2) = \sqrt[3]{5}\,\sqrt[3]{25} - 2\sqrt[3]{5} + 7\sqrt[3]{25} - 7\cdot 2$

$$= \sqrt[3]{125} - 2\sqrt[3]{5} + 7\sqrt[3]{25} - 14$$
$$= 5 - 2\sqrt[3]{5} + 7\sqrt[3]{25} - 14$$
$$= -9 - 2\sqrt[3]{5} + 7\sqrt[3]{25}$$

73. $(3\sqrt[3]{9} + 1)^2 = (3\sqrt[3]{9})^2 + 2\cdot 3\sqrt[3]{9}\cdot 1 + 1^2 = 9\sqrt[3]{81} + 6\sqrt[3]{9} + 1$

$$= 9\sqrt[3]{27}\,\sqrt[3]{3} + 6\sqrt[3]{9} + 1$$
$$= 9\cdot 3\sqrt[3]{3} + 6\sqrt[3]{9} + 1$$
$$= 27\sqrt[3]{3} + 6\sqrt[3]{9} + 1$$

75. $(\sqrt[3]{12} - \sqrt[3]{2})(\sqrt[3]{12} + \sqrt[3]{2}) = (\sqrt[3]{12})^2 - (\sqrt[3]{2})^2 = \sqrt[3]{144} - \sqrt[3]{4}$

$$= \sqrt[3]{8}\,\sqrt[3]{18} - \sqrt[3]{4}$$
$$= 2\sqrt[3]{18} - \sqrt[3]{4}$$

77. $P = 2\ell + 2w$

$P = 2(\sqrt{5} + 8) + 2(\sqrt{5} + 3) = 2\sqrt{5} + 16 + 2\sqrt{5} + 6 = 4\sqrt{5} + 22$

$A = \ell w$

$A = (\sqrt{5} + 8)(\sqrt{5} + 3) = \sqrt{5}\,\sqrt{5} + 3\sqrt{5} + 8\sqrt{5} + 8\cdot 3 = 5 + 11\sqrt{5} + 24$

$$= 29 + 11\sqrt{5}$$

79. $c = \sqrt{a^2 + b^2}$

$x = \sqrt{(\sqrt{2})^2 + (\sqrt{2})^2} = \sqrt{2 + 2} = \sqrt{4} = 2$

81. $a = \sqrt{c^2 - b^2}$

$x = \sqrt{(3\sqrt{2})^2 - (\sqrt{6})^2} = \sqrt{9 \cdot 2 - 6} = \sqrt{12} = \sqrt{4}\sqrt{3} = 2\sqrt{3}$

Problem Set 8.6, pp. 389-391

1. $\dfrac{1}{\sqrt{3}} = \dfrac{1}{\sqrt{3}} \cdot \dfrac{\sqrt{3}}{\sqrt{3}} = \dfrac{\sqrt{3}}{3}$

3. $\dfrac{15}{\sqrt{6}} = \dfrac{15}{\sqrt{6}} \cdot \dfrac{\sqrt{6}}{\sqrt{6}} = \dfrac{15\sqrt{6}}{6} = \dfrac{5\sqrt{6}}{2}$

5. $\dfrac{2}{\sqrt{2}} = \dfrac{2}{\sqrt{2}} \cdot \dfrac{\sqrt{2}}{\sqrt{2}} = \dfrac{2\sqrt{2}}{2} = \sqrt{2}$

7. $\sqrt{\dfrac{5}{3}} = \dfrac{\sqrt{5}}{\sqrt{3}} = \dfrac{\sqrt{5}}{\sqrt{3}} \cdot \dfrac{\sqrt{3}}{\sqrt{3}} = \dfrac{\sqrt{15}}{3}$

9. $\dfrac{\sqrt{10}}{\sqrt{2}} = \sqrt{\dfrac{10}{2}} = \sqrt{5}$

11. $\dfrac{\sqrt{27}}{\sqrt{3}} = \sqrt{\dfrac{27}{3}} = \sqrt{9} = 3$

13. $\sqrt{\dfrac{20}{7}} = \dfrac{\sqrt{20}}{\sqrt{7}} = \dfrac{\sqrt{4}\sqrt{5}}{\sqrt{7}} = \dfrac{2\sqrt{5}}{\sqrt{7}} = \dfrac{2\sqrt{5}}{\sqrt{7}} \cdot \dfrac{\sqrt{7}}{\sqrt{7}} = \dfrac{2\sqrt{35}}{7}$

15. $\dfrac{15}{2\sqrt{10}} = \dfrac{15}{2\sqrt{10}} \cdot \dfrac{\sqrt{10}}{\sqrt{10}} = \dfrac{15\sqrt{10}}{2 \cdot 10} = \dfrac{15\sqrt{10}}{20} = \dfrac{3\sqrt{10}}{4}$

17. $\sqrt{\dfrac{1}{8}} = \dfrac{\sqrt{1}}{\sqrt{8}} = \dfrac{1}{\sqrt{4}\sqrt{2}} = \dfrac{1}{2\sqrt{2}} = \dfrac{1}{2\sqrt{2}} \cdot \dfrac{\sqrt{2}}{\sqrt{2}} = \dfrac{\sqrt{2}}{2 \cdot 2} = \dfrac{\sqrt{2}}{4}$

19. $\dfrac{12\sqrt{3}}{8\sqrt{11}} = \dfrac{3\sqrt{3}}{2\sqrt{11}} = \dfrac{3\sqrt{3}}{2\sqrt{11}} \cdot \dfrac{\sqrt{11}}{\sqrt{11}} = \dfrac{3\sqrt{33}}{2 \cdot 11} = \dfrac{3\sqrt{33}}{22}$

21. $\dfrac{10}{\sqrt{75}} = \dfrac{10}{\sqrt{25}\sqrt{3}} = \dfrac{10}{5\sqrt{3}} = \dfrac{2}{\sqrt{3}} = \dfrac{2}{\sqrt{3}} \cdot \dfrac{\sqrt{3}}{\sqrt{3}} = \dfrac{2\sqrt{3}}{3}$

23. $\dfrac{\sqrt{4x}}{\sqrt{y}} = \dfrac{\sqrt{4}\,\sqrt{x}}{\sqrt{y}} = \dfrac{2\sqrt{x}}{\sqrt{y}} = \dfrac{2\sqrt{x}}{\sqrt{y}} \cdot \dfrac{\sqrt{y}}{\sqrt{y}} = \dfrac{2\sqrt{xy}}{y}$

25. $\dfrac{4}{\sqrt{6x}} = \dfrac{4}{\sqrt{6x}} \cdot \dfrac{\sqrt{6x}}{\sqrt{6x}} = \dfrac{4\sqrt{6x}}{6x} = \dfrac{2\sqrt{6x}}{3x}$

27. $\sqrt{\dfrac{a^3}{9b}} = \dfrac{\sqrt{a^3}}{\sqrt{9b}} = \dfrac{\sqrt{a^2}\,\sqrt{a}}{\sqrt{9}\,\sqrt{b}} = \dfrac{a\sqrt{a}}{3\sqrt{b}} = \dfrac{a\sqrt{a}}{3\sqrt{b}} \cdot \dfrac{\sqrt{b}}{\sqrt{b}} = \dfrac{a\sqrt{ab}}{3b}$

29. $\dfrac{14\sqrt{5x}}{\sqrt{32y^2}} = \dfrac{14\sqrt{5x}}{\sqrt{16y^2}\,\sqrt{2}} = \dfrac{14\sqrt{5x}}{4y\sqrt{2}} = \dfrac{7\sqrt{5x}}{2y\sqrt{2}} = \dfrac{7\sqrt{5x}}{2y\sqrt{2}} \cdot \dfrac{\sqrt{2}}{\sqrt{2}} = \dfrac{7\sqrt{10x}}{2y\cdot 2} = \dfrac{7\sqrt{10x}}{4y}$

31. $\dfrac{\sqrt{11x^7}}{\sqrt{77xy}} = \sqrt{\dfrac{11x^7}{77xy}} = \sqrt{\dfrac{x^6}{7y}} = \dfrac{\sqrt{x^6}}{\sqrt{7y}} = \dfrac{x^3}{\sqrt{7y}} = \dfrac{x^3}{\sqrt{7y}} \cdot \dfrac{\sqrt{7y}}{\sqrt{7y}} = \dfrac{x^3\sqrt{7y}}{7y}$

33. $\sqrt{\dfrac{2}{5}} \cdot \sqrt{\dfrac{2}{9}} = \sqrt{\dfrac{4}{45}} = \dfrac{\sqrt{4}}{\sqrt{45}} = \dfrac{2}{\sqrt{9}\,\sqrt{5}} = \dfrac{2}{3\sqrt{5}} = \dfrac{2}{3\sqrt{5}} \cdot \dfrac{\sqrt{5}}{\sqrt{5}} = \dfrac{2\sqrt{5}}{3\cdot 5} = \dfrac{2\sqrt{5}}{15}$

35. $\sqrt{\dfrac{7}{12}} \cdot \sqrt{\dfrac{1}{6}} = \sqrt{\dfrac{7}{72}} = \dfrac{\sqrt{7}}{\sqrt{36}\,\sqrt{2}} = \dfrac{\sqrt{7}}{6\sqrt{2}} = \dfrac{\sqrt{7}}{6\sqrt{2}} \cdot \dfrac{\sqrt{2}}{\sqrt{2}} = \dfrac{\sqrt{14}}{6\cdot 2} = \dfrac{\sqrt{14}}{12}$

37. $\dfrac{1}{\sqrt[3]{3}} = \dfrac{1}{\sqrt[3]{3}} \cdot \dfrac{\sqrt[3]{9}}{\sqrt[3]{9}} = \dfrac{\sqrt[3]{9}}{\sqrt[3]{27}} = \dfrac{\sqrt[3]{9}}{3}$

39. $\dfrac{6}{\sqrt[3]{2}} = \dfrac{6}{\sqrt[3]{2}} \cdot \dfrac{\sqrt[3]{4}}{\sqrt[3]{4}} = \dfrac{6\sqrt[3]{4}}{\sqrt[3]{8}} = \dfrac{6\sqrt[3]{4}}{2} = 3\sqrt[3]{4}$

41. $\sqrt[3]{\dfrac{3}{4}} = \dfrac{\sqrt[3]{3}}{\sqrt[3]{4}} = \dfrac{\sqrt[3]{3}}{\sqrt[3]{4}} \cdot \dfrac{\sqrt[3]{2}}{\sqrt[3]{2}} = \dfrac{\sqrt[3]{6}}{\sqrt[3]{8}} = \dfrac{\sqrt[3]{6}}{2}$

43. $\sqrt[3]{\dfrac{8}{25}} = \dfrac{\sqrt[3]{8}}{\sqrt[3]{25}} = \dfrac{2}{\sqrt[3]{25}} = \dfrac{2}{\sqrt[3]{25}} \cdot \dfrac{\sqrt[3]{5}}{\sqrt[3]{5}} = \dfrac{2\sqrt[3]{5}}{\sqrt[3]{125}} = \dfrac{2\sqrt[3]{5}}{5}$

45. $\sqrt[3]{\dfrac{1}{72}} = \dfrac{\sqrt[3]{1}}{\sqrt[3]{72}} = \dfrac{1}{\sqrt[3]{8}\,\sqrt[3]{9}} = \dfrac{1}{2\sqrt[3]{9}} = \dfrac{1}{2\sqrt[3]{9}} \cdot \dfrac{\sqrt[3]{3}}{\sqrt[3]{3}} = \dfrac{\sqrt[3]{3}}{2\sqrt[3]{27}} = \dfrac{\sqrt[3]{3}}{2\cdot 3} = \dfrac{\sqrt[3]{3}}{6}$

47. $\sqrt[3]{\dfrac{5}{6x^2}} = \dfrac{\sqrt[3]{5}}{\sqrt[3]{6x^2}} = \dfrac{\sqrt[3]{5}}{\sqrt[3]{6x^2}} \cdot \dfrac{\sqrt[3]{36x}}{\sqrt[3]{36x}} = \dfrac{\sqrt[3]{180x}}{\sqrt[3]{216x^3}} = \dfrac{\sqrt[3]{180x}}{6x}$

49. $\dfrac{1}{\sqrt{7} - 1} = \dfrac{1}{\sqrt{7} - 1} \cdot \dfrac{\sqrt{7} + 1}{\sqrt{7} + 1} = \dfrac{\sqrt{7} + 1}{(\sqrt{7})^2 - 1^2} = \dfrac{\sqrt{7} + 1}{7 - 1} = \dfrac{\sqrt{7} + 1}{6}$

51. $\dfrac{2}{-3 - \sqrt{7}} = \dfrac{2}{-3 - \sqrt{7}} \cdot \dfrac{-3 + \sqrt{7}}{-3 + \sqrt{7}} = \dfrac{2(-3 + \sqrt{7})}{(-3)^2 - (\sqrt{7})^2} = \dfrac{2(-3 + \sqrt{7})}{9 - 7} = \dfrac{2(-3 + \sqrt{7})}{2}$

$$= -3 + \sqrt{7}$$

53. $\dfrac{3}{3 + \sqrt{3}} = \dfrac{3}{3 + \sqrt{3}} \cdot \dfrac{3 - \sqrt{3}}{3 - \sqrt{3}} = \dfrac{3(3 - \sqrt{3})}{3^2 - (\sqrt{3})^2} = \dfrac{3(3 - \sqrt{3})}{9 - 3} = \dfrac{3(3 - \sqrt{3})}{6}$

$$= \dfrac{3 - \sqrt{3}}{2}$$

55. $\dfrac{\sqrt{5}}{\sqrt{5} - \sqrt{2}} = \dfrac{\sqrt{5}}{\sqrt{5} - \sqrt{2}} \cdot \dfrac{\sqrt{5} + \sqrt{2}}{\sqrt{5} + \sqrt{2}} = \dfrac{\sqrt{5}\sqrt{5} + \sqrt{5}\sqrt{2}}{(\sqrt{5})^2 - (\sqrt{2})^2} = \dfrac{5 + \sqrt{10}}{5 - 2} = \dfrac{5 + \sqrt{10}}{3}$

57. $\dfrac{4\sqrt{3}}{\sqrt{7} + \sqrt{5}} = \dfrac{4\sqrt{3}}{\sqrt{7} + \sqrt{5}} \cdot \dfrac{\sqrt{7} - \sqrt{5}}{\sqrt{7} - \sqrt{5}} = \dfrac{4\sqrt{3}(\sqrt{7} - \sqrt{5})}{(\sqrt{7})^2 - (\sqrt{5})^2} = \dfrac{4\sqrt{3}(\sqrt{7} - \sqrt{5})}{7 - 5}$

$$= \dfrac{4\sqrt{3}(\sqrt{7} - \sqrt{5})}{2}$$

$$= 2\sqrt{3}(\sqrt{7} - \sqrt{5})$$

$$= 2\sqrt{21} - 2\sqrt{15}$$

59. $\dfrac{\sqrt{5} - \sqrt{6}}{\sqrt{5} + \sqrt{6}} = \dfrac{\sqrt{5} - \sqrt{6}}{\sqrt{5} + \sqrt{6}} \cdot \dfrac{\sqrt{5} - \sqrt{6}}{\sqrt{5} - \sqrt{6}} = \dfrac{(\sqrt{5})^2 - 2\sqrt{5}\sqrt{6} + (\sqrt{6})^2}{(\sqrt{5})^2 - (\sqrt{6})^2}$

$$= \dfrac{5 - 2\sqrt{30} + 6}{5 - 6}$$

$$= \dfrac{11 - 2\sqrt{30}}{-1}$$

$$= -11 + 2\sqrt{30}$$

61. $\dfrac{\sqrt{2} + 1}{-\sqrt{5} + \sqrt{3}} = \dfrac{\sqrt{2} + 1}{-\sqrt{5} + \sqrt{3}} \cdot \dfrac{-\sqrt{5} - \sqrt{3}}{-\sqrt{5} - \sqrt{3}} = \dfrac{-\sqrt{10} - \sqrt{6} - \sqrt{5} - \sqrt{3}}{(-\sqrt{5})^2 - (\sqrt{3})^2}$

$$= \dfrac{-\sqrt{10} - \sqrt{6} - \sqrt{5} - \sqrt{3}}{5 - 3}$$

$$= \dfrac{-\sqrt{10} - \sqrt{6} - \sqrt{5} - \sqrt{3}}{2}$$

63. $\dfrac{4 + \sqrt{6}}{\sqrt{3} - \sqrt{2}} = \dfrac{4 + \sqrt{6}}{\sqrt{3} - \sqrt{2}} \cdot \dfrac{\sqrt{3} + \sqrt{2}}{\sqrt{3} + \sqrt{2}} = \dfrac{4\sqrt{3} + 4\sqrt{2} + \sqrt{18} + \sqrt{12}}{(\sqrt{3})^2 - (\sqrt{2})^2}$

$$= \dfrac{4\sqrt{3} + 4\sqrt{2} + \sqrt{9}\sqrt{2} + \sqrt{4}\sqrt{3}}{3 - 2}$$

$$= \dfrac{4\sqrt{3} + 4\sqrt{2} + 3\sqrt{2} + 2\sqrt{3}}{1}$$

$$= 6\sqrt{3} + 7\sqrt{2}$$

65. $\dfrac{5\sqrt{14} - 3\sqrt{10}}{2\sqrt{2}} = \dfrac{5\sqrt{14} - 3\sqrt{10}}{2\sqrt{2}} \cdot \dfrac{\sqrt{2}}{\sqrt{2}} = \dfrac{5\sqrt{28} - 3\sqrt{20}}{2 \cdot 2}$

$$= \dfrac{5\sqrt{4}\sqrt{7} - 3\sqrt{4}\sqrt{5}}{4}$$

$$= \dfrac{10\sqrt{7} - 6\sqrt{5}}{4}$$

$$= \dfrac{2(5\sqrt{7} - 3\sqrt{5})}{4}$$

$$= \dfrac{5\sqrt{7} - 3\sqrt{5}}{2}$$

67. $\dfrac{\sqrt{2}}{3\sqrt{3} - 2\sqrt{2}} = \dfrac{\sqrt{2}}{3\sqrt{3} - 2\sqrt{2}} \cdot \dfrac{3\sqrt{3} + 2\sqrt{2}}{3\sqrt{3} + 2\sqrt{2}} = \dfrac{3\sqrt{6} + 2\cdot 2}{(3\sqrt{3})^2 - (2\sqrt{2})^2}$

$$= \dfrac{3\sqrt{6} + 4}{9\cdot 3 - 4\cdot 2}$$

$$= \dfrac{3\sqrt{6} + 4}{19}$$

69. $\dfrac{\sqrt{x} + \sqrt{y}}{\sqrt{x} - \sqrt{y}} = \dfrac{\sqrt{x} + \sqrt{y}}{\sqrt{x} - \sqrt{y}} \cdot \dfrac{\sqrt{x} + \sqrt{y}}{\sqrt{x} + \sqrt{y}} = \dfrac{(\sqrt{x})^2 + 2\sqrt{x}\sqrt{y} + (\sqrt{y})^2}{(\sqrt{x})^2 - (\sqrt{y})^2}$

$$= \dfrac{x + 2\sqrt{xy} + y}{x - y}$$

71. $\dfrac{\sqrt{2x}}{\sqrt{2x} + 3} = \dfrac{\sqrt{2x}}{\sqrt{2x} + 3} \cdot \dfrac{\sqrt{2x} - 3}{\sqrt{2x} - 3} = \dfrac{\sqrt{2x}\sqrt{2x} - \sqrt{2x}\cdot 3}{(\sqrt{2x})^2 - 3^2} = \dfrac{2x - 3\sqrt{2x}}{2x - 9}$

73. $A = \sqrt{\dfrac{W}{R}}$

$A = \sqrt{\dfrac{1100}{24}} = \sqrt{\dfrac{275}{6}} = \dfrac{\sqrt{275}}{\sqrt{6}} = \dfrac{\sqrt{25}\sqrt{11}}{\sqrt{6}} = \dfrac{5\sqrt{11}}{\sqrt{6}} = \dfrac{5\sqrt{11}}{\sqrt{6}} \cdot \dfrac{\sqrt{6}}{\sqrt{6}} = \dfrac{5\sqrt{66}}{6}$ amps

Problem Set 8.7, pp. 395-396

1. $\sqrt{x} = 2$

   $(\sqrt{x})^2 = 2^2$

   $x = 4$

3. $\sqrt{x + 1} = 3$

   $(\sqrt{x + 1})^2 = 3^2$

   $x + 1 = 9$

   $x = 8$

5.
$$\sqrt{2y - 1} = 0$$
$$(\sqrt{2y - 1})^2 = 0^2$$
$$2y - 1 = 0$$
$$2y = 1$$
$$y = \frac{1}{2}$$

7. $\sqrt{m} - 1 = 5$
$$\sqrt{m} = 6$$
$$(\sqrt{m})^2 = 6^2$$
$$m = 36$$

9. $\sqrt{3p - 5} + 7 = 3$
$$\sqrt{3p - 5} = -4$$
$$(\sqrt{3p - 5})^2 = (-4)^2$$
$$3p - 5 = 16$$
$$3p = 21$$
$$p = 7$$

Since 7 does not check, there is no solution.

11. $\sqrt{5r + 1} + 3 = 9$
$$\sqrt{5r + 1} = 6$$
$$(\sqrt{5r + 1})^2 = 6^2$$
$$5r + 1 = 36$$
$$5r = 35$$
$$r = 7$$

13.
$$\sqrt{3t - 1} = \sqrt{t + 3}$$
$$(\sqrt{3t - 1})^2 = (\sqrt{t + 3})^2$$
$$3t - 1 = t + 3$$
$$2t = 4$$
$$t = 2$$

15.
$$\sqrt{4x + 5} = \sqrt{3x + 4}$$
$$(\sqrt{4x + 5})^2 = (\sqrt{3x + 4})^2$$
$$4x + 5 = 3x + 4$$
$$x = -1$$

17.
$$3\sqrt{m} = \sqrt{m + 8}$$
$$(3\sqrt{m})^2 = (\sqrt{m + 8})^2$$
$$9m = m + 8$$
$$8m = 8$$
$$m = 1$$

19.
$$2\sqrt{y} = \sqrt{y + 9}$$
$$(2\sqrt{y})^2 = (\sqrt{y + 9})^2$$
$$4y = y + 9$$
$$3y = 9$$
$$y = 3$$

21.

$$\sqrt{p + 2} = p$$

$$(\sqrt{p + 2})^2 = p^2$$

$$p + 2 = p^2$$

$$0 = p^2 - p - 2$$

$$0 = (p - 2)(p + 1)$$

$$p - 2 = 0 \quad \text{or} \quad p + 1 = 0$$

$$p = 2 \qquad\qquad p = -1$$

Since -1 does not check, the only solution is 2.

23.

$$\sqrt{r^2 - 3} - 1 = 0$$

$$\sqrt{r^2 - 3} = 1$$

$$(\sqrt{r^2 - 3})^2 = 1^2$$

$$r^2 - 3 = 1$$

$$r^2 - 4 = 0$$

$$(r - 2)(r + 2) = 0$$

$$r - 2 = 0 \quad \text{or} \quad r + 2 = 0$$

$$r = 2 \qquad\qquad r = -2$$

25.

$$\sqrt{x^2 - 8x + 19} = 2$$

$$(\sqrt{x^2 - 8x + 19})^2 = 2^2$$

$$x^2 - 8x + 19 = 4$$

$$x^2 - 8x + 15 = 0$$

$$(x - 3)(x - 5) = 0$$

$$x - 3 = 0 \quad \text{or} \quad x - 5 = 0$$

$$x = 3 \qquad\qquad x = 5$$

27.

$$\sqrt{y^2 + 9y + 45} = y$$

$$(\sqrt{y^2 + 9y + 45})^2 = y^2$$

$$y^2 + 9y + 45 = y^2$$

$$9y + 45 = 0$$

$$9y = -45$$

$$y = -5$$

Since -5 does not check, there is no solution.

29.

$$\sqrt{4z^2 + 2z - 1} = z$$

$$(\sqrt{4z^2 + 2z - 1})^2 = z^2$$

$$4z^2 + 2z - 1 = z^2$$

$$3z^2 + 2z - 1 = 0$$

$$(3z - 1)(z + 1) = 0$$

$$3z - 1 = 0 \quad \text{or} \quad z + 1 = 0$$

$$3z = 1 \qquad\qquad z = -1$$

$$z = \frac{1}{3}$$

Since -1 does not check, the only solution is 1/3.

31.

$$\sqrt{x^2 + 5} = x + 5$$

$$(\sqrt{x^2 + 5})^2 = (x + 5)^2$$

$$x^2 + 5 = x^2 + 10x + 25$$

$$5 = 10x + 25$$

$$-20 = 10x$$

$$-2 = x$$

33. $\sqrt{y + 4} = y + 2$

$(\sqrt{y + 4})^2 = (y + 2)^2$

$y + 4 = y^2 + 4y + 4$

$0 = y^2 + 3y$

$0 = y(y + 3)$

$y = 0$ or $y + 3 = 0$

$y = -3$

Since -3 does not check, the only solution is 0.

37. $\sqrt{r} + 2 = r$

$\sqrt{r} = r - 2$

$(\sqrt{r})^2 = (r - 2)^2$

$r = r^2 - 4r + 4$

$0 = r^2 - 5r + 4$

$0 = (r - 4)(r - 1)$

$r - 4 = 0$ or $r - 1 = 0$

$r = 4$ $r = 1$

Since 1 does not check, the only solution is 4.

41. $2\sqrt{r + 3} = r + 4$

$(2\sqrt{r + 3})^2 = (r + 4)^2$

$4(r + 3) = r^2 + 8r + 16$

$4r + 12 = r^2 + 8r + 16$

$0 = r^2 + 4r + 4$

$0 = (r + 2)(r + 2)$

$r + 2 = 0$ or $r + 2 = 0$

$r = -2$ $r = -2$

35. $\sqrt{2m - 3} = m - 3$

$(\sqrt{2m - 3})^2 = (m - 3)^2$

$2m - 3 = m^2 - 6m + 9$

$0 = m^2 - 8m + 12$

$0 = (m - 2)(m - 6)$

$m - 2 = 0$ or $m - 6 = 0$

$m = 2$ $m = 6$

Since 2 does not check, the only solution is 6.

39. $\sqrt{p^2 + 9} + 1 = p$

$\sqrt{p^2 + 9} = p - 1$

$(\sqrt{p^2 + 9})^2 = (p - 1)^2$

$p^2 + 9 = p^2 - 2p + 1$

$8 = -2p$

$-4 = p$

Since -4 does not check, there is no solution.

43. $\sqrt{x - 17} = \sqrt{17 - x}$

$(\sqrt{x - 17})^2 = (\sqrt{17 - x})^2$

$x - 17 = 17 - x$

$2x = 34$

$x = 17$

45.
$$x + 1 = \sqrt{x + 1}$$
$$(x + 1)^2 = (\sqrt{x + 1})^2$$
$$x^2 + 2x + 1 = x + 1$$
$$x^2 + x = 0$$
$$x(x + 1) = 0$$
$$x = 0 \quad \text{or} \quad x + 1 = 0$$
$$x = -1$$

47.
$$3 = \sqrt{y - 3} + y$$
$$3 - y = \sqrt{y - 3}$$
$$(3 - y)^2 = (\sqrt{y - 3})^2$$
$$9 - 6y + y^2 = y - 3$$
$$y^2 - 7y + 12 = 0$$
$$(y - 3)(y - 4) = 0$$
$$y - 3 = 0 \quad \text{or} \quad y - 4 = 0$$
$$y = 3 \qquad y = 4$$

Since 4 does not check, the only solution is 3.

49.
$$\sqrt{4z + 5} - 5 = 2z - 10$$
$$\sqrt{4z + 5} = 2z - 5$$
$$(\sqrt{4z + 5})^2 = (2z - 5)^2$$
$$4z + 5 = 4z^2 - 20z + 25$$
$$0 = 4z^2 - 24z + 20$$
$$0 = z^2 - 6z + 5$$
$$0 = (z - 5)(z - 1)$$
$$z - 5 = 0 \quad \text{or} \quad z - 1 = 0$$
$$z = 5 \qquad z = 1$$

Since 1 does not check, the only solution is 5.

51.
$$\sqrt{t^2 - 7t - 14} = t - 6$$
$$(\sqrt{t^2 - 7t - 14})^2 = (t - 6)^2$$
$$t^2 - 7t - 14 = t^2 - 12t + 36$$
$$-7t - 14 = -12t + 36$$
$$5t = 50$$
$$t = 10$$

53. $\sqrt{10k^2 + 2k - 7} = 3k + 2$

$(\sqrt{10k^2 + 2k - 7})^2 = (3k + 2)^2$

$10k^2 + 2k - 7 = 9k^2 + 12k + 4$

$k^2 - 10k - 11 = 0$

$(k - 11)(k + 1) = 0$

$k - 11 = 0 \quad or \quad k + 1 = 0$

$k = 11 \qquad k = -1$

Since -1 does not check, the only solution is 11.

55. $\sqrt{4 + 3\sqrt{x}} = 5$

$(\sqrt{4 + 3\sqrt{x}})^2 = 5^2$

$4 + 3\sqrt{x} = 25$

$3\sqrt{x} = 21$

$\sqrt{x} = 7$

$(\sqrt{x})^2 = 7^2$

$x = 49$

57. $\sqrt{x + 7} - \sqrt{x} = 1$

$\sqrt{x + 7} = 1 + \sqrt{x}$

$(\sqrt{x + 7})^2 = (1 + \sqrt{x})^2$

$x + 7 = 1 + 2\sqrt{x} + x$

$6 = 2\sqrt{x}$

$3 = \sqrt{x}$

$3^2 = (\sqrt{x})^2$

$9 = x$

59. $\sqrt{3y + 4} - \sqrt{y} = 2$

$\sqrt{3y + 4} = 2 + \sqrt{y}$

$(\sqrt{3y + 4})^2 = (2 + \sqrt{y})^2$

$3y + 4 = 4 + 4\sqrt{y} + y$

$2y = 4\sqrt{y}$

$y = 2\sqrt{y}$

$y^2 = (2\sqrt{y})^2$

$y^2 = 4y$

$y^2 - 4y = 0$

$y(y - 4) = 0$

$y = 0 \quad or \quad y - 4 = 0$

$y = 4$

61. $t = \dfrac{\sqrt{d}}{4}$

$4 = \dfrac{\sqrt{d}}{4}$

$16 = \sqrt{d}$

$16^2 = (\sqrt{d})^2$

$d = 256 \text{ ft}$

63. x = the number

$\sqrt{x + 4} = 3$

$(\sqrt{x + 4})^2 = 3^2$

$x + 4 = 9$

$x = 5$

65. x = the number

$2\sqrt{x} = x - 3$

$(2\sqrt{x})^2 = (x - 3)^2$

$4x = x^2 - 6x + 9$

$0 = x^2 - 10x + 9$

$0 = (x - 1)(x - 9)$

$x - 1 = 0 \quad \text{or} \quad x - 9 = 0$

$x = 1 \qquad\qquad x = 9$

Since 1 does not check, the only solution is 9.

67. $V = \sqrt{WR}$

$V^2 = (\sqrt{WR})^2$

$V^2 = WR$

$\dfrac{V^2}{W} = \dfrac{WR}{W}$

$R = \dfrac{V^2}{W}$

Problem Set 8.8, pp. 400-401

1. $16^{1/2} = \sqrt{16} = 4$

3. $25^{1/2} = \sqrt{25} = 5$

5. $8^{1/3} = \sqrt[3]{8} = 2$

7. $64^{1/3} = \sqrt[3]{64} = 4$

9. $81^{1/4} = \sqrt[4]{81} = 3$

11. $27^{2/3} = (\sqrt[3]{27})^2 = 3^2 = 9$

13. $9^{3/2} = (\sqrt{9})^3 = 3^3 = 27$

15. $32^{3/5} = (\sqrt[5]{32})^3 = 2^3 = 8$

17. $121^{-1/2} = \dfrac{1}{121^{1/2}} = \dfrac{1}{\sqrt{121}} = \dfrac{1}{11}$

19. $16^{-5/4} = \dfrac{1}{16^{5/4}} = \dfrac{1}{(\sqrt[4]{16})^5} = \dfrac{1}{2^5} = \dfrac{1}{32}$

21. $(-8)^{2/3} = (\sqrt[3]{-8})^2 = (-2)^2 = 4$

23. $(-125)^{-1/3} = \dfrac{1}{(-125)^{1/3}} = \dfrac{1}{\sqrt[3]{-125}} = \dfrac{1}{-5} = -\dfrac{1}{5}$

25. $2^{2/3} \cdot 2^{5/3} = 2^{2/3\,+\,5/3} = 2^{7/3}$

27. $5^{3/2} \cdot 5^{-1} = 5^{3/2\,+\,(-1)} = 5^{3/2\,+\,(-2/2)} = 5^{1/2}$

29. $\dfrac{11^{4/5}}{11^{1/5}} = 11^{4/5\,-\,1/5} = 11^{3/5}$

31. $\dfrac{12}{12^{1/3}} = 12^{1\,-\,1/3} = 12^{3/3\,-\,1/3} = 12^{2/3}$

33. $(10^{1/4})^2 = 10^{2(1/4)} = 10^{1/2}$

35. $(3^{2/5})^{10} = 3^{10(2/5)} = 3^4$

37. $(2^{1/3} \cdot 6^{1/3})^3 = (2^{1/3})^3 (6^{1/3})^3 = 2^{3(1/3)} 6^{3(1/3)} = 2^1 \cdot 6^1 = 12$

39. $(9^{2/5} \cdot 3^{-1/5})^5 = (9^{2/5})^5 (3^{-1/5})^5 = 9^{5(2/5)} 3^{5(-1/5)} = 9^2 \cdot 3^{-1}$

$$= 81 \cdot \dfrac{1}{3}$$

$$= 27$$

41. $(\dfrac{9}{16})^{3/2} = (\sqrt{\dfrac{9}{16}})^3 = (\dfrac{3}{4})^3 = \dfrac{27}{64}$

43. $(\frac{8}{27})^{-1/3} = (\frac{27}{8})^{1/3} = \sqrt[3]{\frac{27}{8}} = \frac{3}{2}$

45. $x^{1/3} \cdot x^{-1/3} = x^{1/3 + (-1/3)} = x^0 = 1$

47. $m^{1/2} \cdot m^{3/2} = m^{1/2 + 3/2} = m^{4/2} = m^2$

49. $\frac{a^{5/2}}{a^{3/2}} = a^{5/2 - 3/2} = a^{2/2} = a^1 = a$

51. $\frac{y^{1/6}}{y^{3/6}} = y^{1/6 - 3/6} = y^{-2/6} = y^{-1/3} = \frac{1}{y^{1/3}}$

53. $\frac{z^{-1/4}}{z^{3/4}} = z^{-1/4 - 3/4} = z^{-4/4} = z^{-1} = \frac{1}{z}$

55. $(p^6)^{1/2} = p^{6(1/2)} = p^3$

57. $(r^{2/3})^2 = r^{2(2/3)} = r^{4/3}$

59. $(27m^6)^{1/3} = 27^{1/3}(m^6)^{1/3} = \sqrt[3]{27} \; m^{6(1/3)} = 3m^2$

61. $(a^{-2}b^{1/5})^5 = (a^{-2})^5(b^{1/5})^5 = a^{(-2)5}b^{5(1/5)} = a^{-10}b^1 = \frac{1}{a^{10}} \cdot b = \frac{b}{a^{10}}$

63. $(\frac{c^2}{d^2})^{3/2} = \frac{(c^2)^{3/2}}{(d^2)^{3/2}} = \frac{c^{2(3/2)}}{d^{2(3/2)}} = \frac{c^3}{d^3}$

65. $(\frac{x^{3/4}}{y^{3/2}})^{2/3} = \frac{(x^{3/4})^{2/3}}{(y^{3/2})^{2/3}} = \frac{x^{(3/4)(2/3)}}{y^{(3/2)(2/3)}} = \frac{x^{1/2}}{y}$

67. $\dfrac{k \cdot k^{-5/2}}{k^{1/4} \cdot k^{-3/4}} = \dfrac{k^{1 + (-5/2)}}{k^{1/4 + (-3/4)}} = \dfrac{k^{2/2 + (-5/2)}}{k^{-2/4}} = \dfrac{k^{-3/2}}{k^{-1/2}} = k^{-3/2 - (-1/2)}$

$$= k^{-3/2 + 1/2}$$

$$= k^{-2/2}$$

$$= k^{-1}$$

$$= \frac{1}{k}$$

69. $\sqrt[4]{4} = \sqrt[4]{2^2} = 2^{2/4} = 2^{1/2} = \sqrt{2}$

71. $\sqrt[6]{36^3} = \sqrt[6]{(6^2)^3} = \sqrt[6]{6^6} = 6^{6/6} = 6$

73. $\sqrt[8]{r^2} = r^{2/8} = r^{1/4} = \sqrt[4]{r}$

75. $\sqrt[12]{x^8} = x^{8/12} = x^{2/3} = \sqrt[3]{x^2}$

77. $\sqrt{\sqrt{2}} = \sqrt{2^{1/2}} = (2^{1/2})^{1/2} = 2^{(1/2)(1/2)} = 2^{1/4} = \sqrt[4]{2}$

79. $\sqrt[3]{\sqrt{y}} = \sqrt[3]{y^{1/2}} = (y^{1/2})^{1/3} = y^{(1/2)(1/3)} = y^{1/6} = \sqrt[6]{y}$

81. $R = \dfrac{A}{D^{2/3}}$

$R = \dfrac{320}{64^{2/3}} = \dfrac{320}{(\sqrt[3]{64})^2} = \dfrac{320}{4^2} = \dfrac{320}{16} = 20$

83. $\boxed{\text{Clear}}$ 81 $\boxed{\sqrt{x}}$ 9

$81^{1/2} = 9$

85. $\boxed{\text{Clear}}$ 64 $\boxed{\text{INV}}$ $\boxed{y^x}$ 3 $\boxed{y^x}$ 2 $\boxed{=}$ 16

    $64^{2/3} = 16$

87. $\boxed{\text{Clear}}$ 3 $\boxed{\sqrt{x}}$ $\boxed{\sqrt{x}}$ 1.3160740

    $3^{1/4} \approx 1.32$

89. $\boxed{\text{Clear}}$ 1024 $\boxed{+/-}$ $\boxed{\text{INV}}$ $\boxed{y^x}$ 5 $\boxed{=}$ -4

    $(-1024)^{1/5} = -4$

91. $y = 0.41d^{3/2}$

    $y = 0.41(36)^{3/2} = 0.41(\sqrt{36})^3 = 0.41(6)^3 = 0.41(216) \approx 89$ days

93. $y = 0.41d^{3/2}$

    $y = 0.41(92.6)^{3/2} \approx 0.41(891.07956) \approx 365$ days

CHAPTER 9

MORE QUADRATIC EQUATIONS

Problem Set 9.1, p. 412

1. (a)
$$x^2 = 25$$
$$x^2 - 25 = 0$$
$$(x - 5)(x + 5) = 0$$
$$x - 5 = 0 \quad \text{or} \quad x + 5 = 0$$
$$x = 5 \qquad x = -5$$

   (b) $x^2 = 25$
$$x = \pm\sqrt{25}$$
$$x = \pm 5$$

3. (a)
$$x^2 - 16 = 0$$
$$(x - 4)(x + 4) = 0$$
$$x - 4 = 0 \quad \text{or} \quad x + 4 = 0$$
$$x = 4 \qquad x = -4$$

   (b) $x^2 - 16 = 0$
$$x^2 = 16$$
$$x = \pm\sqrt{16}$$
$$x = \pm 4$$

5. (a)
$$9x^2 - 4 = 0$$
$$(3x - 2)(3x + 2) = 0$$
$$3x - 2 = 0 \quad \text{or} \quad 3x + 2 = 0$$
$$3x = 2 \qquad\qquad 3x = -2$$
$$x = \frac{2}{3} \qquad\qquad x = -\frac{2}{3}$$

   (b) $9x^2 - 4 = 0$
$$9x^2 = 4$$
$$x^2 = \frac{4}{9}$$
$$x = \pm\sqrt{\frac{4}{9}}$$
$$x = \pm\frac{2}{3}$$

7. (a)
$$x^2 - 2 = 0$$
$$(x - \sqrt{2})(x + \sqrt{2}) = 0$$
$$x - \sqrt{2} = 0 \quad \text{or} \quad x + \sqrt{2} = 0$$
$$x = \sqrt{2} \qquad\qquad x = -\sqrt{2}$$

   (b) $x^2 - 2 = 0$
$$x^2 = 2$$
$$x = \pm\sqrt{2}$$

9. $x^2 - 7 = 0$

$x^2 = 7$

$x = \pm\sqrt{7}$

11. $3y^2 - 60 = 0$

$3y^2 = 60$

$y^2 = 20$

$y = \pm\sqrt{20}$

$y = \pm 2\sqrt{5}$

13. $4m^2 - 45 = 0$

$4m^2 = 45$

$m^2 = \dfrac{45}{4}$

$m = \pm\sqrt{\dfrac{45}{4}}$

$m = \pm\dfrac{\sqrt{45}}{\sqrt{4}}$

$m = \pm\dfrac{3\sqrt{5}}{2}$

15. $p^2 + 9 = 0$

$p^2 = -9$

$p = \pm\sqrt{-9}$

Since $\sqrt{-9}$ is not a real number, there is no real solution.

17. $2r^2 - 1 = 0$

$2r^2 = 1$

$r^2 = \dfrac{1}{2}$

$r = \pm\sqrt{\dfrac{1}{2}}$

$r = \pm\dfrac{\sqrt{1}}{\sqrt{2}}$

$r = \pm\dfrac{1}{\sqrt{2}} \cdot \dfrac{\sqrt{2}}{\sqrt{2}}$

$r = \pm\dfrac{\sqrt{2}}{2}$

19. $(x - 1)^2 = 49$

$x - 1 = \pm\sqrt{49}$

$x - 1 = \pm 7$

$x = 1 \pm 7$

$x = 1 + 7$ or $x = 1 - 7$

$x = 8$ $x = -6$

21. $(p - 3)^2 = 3$

$p - 3 = \pm\sqrt{3}$

$p = 3 \pm \sqrt{3}$

23. $(m + 2)^2 = 18$

$m + 2 = \pm\sqrt{18}$

$m + 2 = \pm 3\sqrt{2}$

$m = -2 \pm 3\sqrt{2}$

25. $(x - 5)^2 = -4$

$x - 5 = \pm\sqrt{-4}$

Since $\sqrt{-4}$ is not a real number, there is no real solution.

27. $(2t + 7)^2 = 9$

$2t + 7 = \pm 3$

$2t = -7 \pm 3$

$t = \dfrac{-7 \pm 3}{2}$

$t = \dfrac{-7 + 3}{2}$ or $t = \dfrac{-7 - 3}{2}$

$t = -2$ $\qquad\qquad$ $t = -5$

29. $(2z - 2)^2 = 1$

$2z - 2 = \pm 1$

$2z = 2 \pm 1$

$z = \dfrac{2 \pm 1}{2}$

$z = \dfrac{2 + 1}{2}$ or $z = \dfrac{2 - 1}{2}$

$z = \dfrac{3}{2}$ $\qquad\qquad$ $z = \dfrac{1}{2}$

31. $(5y + 4)^2 = 11$

$5y + 4 = \pm\sqrt{11}$

$5y = -4 \pm \sqrt{11}$

$y = \dfrac{-4 \pm \sqrt{11}}{5}$

33. $(6r - 3)^2 = 27$

$6r - 3 = \pm\sqrt{27}$

$6r - 3 = \pm 3\sqrt{3}$

$6r = 3 \pm 3\sqrt{3}$

$r = \dfrac{3 \pm 3\sqrt{3}}{6}$

35. $(x - \frac{1}{2})^2 = \frac{9}{4}$

$x - \dfrac{1}{2} = \pm\sqrt{\dfrac{9}{4}}$

$x - \dfrac{1}{2} = \pm\dfrac{3}{2}$

$x = \dfrac{1}{2} \pm \dfrac{3}{2}$

$$r = \frac{3(1 \pm \sqrt{3})}{6}$$

$$r = \frac{1 \pm \sqrt{3}}{2}$$

$$x = \frac{1}{2} + \frac{3}{2} \quad \text{or} \quad x = \frac{1}{2} - \frac{3}{2}$$

$$x = 2 \qquad\qquad x = -1$$

37. $\left(y + \dfrac{3}{2}\right)^2 = \dfrac{5}{4}$

$$y + \frac{3}{2} = \pm\sqrt{\frac{5}{4}}$$

$$y + \frac{3}{2} = \pm\frac{\sqrt{5}}{2}$$

$$y = \frac{-3 \pm \sqrt{5}}{2}$$

39. x = the number

$$5x^2 - 8 = 597$$

$$5x^2 = 605$$

$$x^2 = 121$$

$$x = \pm\sqrt{121}$$

$$x = \pm 11$$

41.   x = width of rectangle

2x = length of rectangle

Area = 200

$$2x \cdot x = 200$$

$$2x^2 = 200$$

$$x^2 = 100$$

$$x = \pm\sqrt{100}$$

$$x = \pm 10$$

The width cannot be -10.
Therefore the width is 10 ft,
and the length is 20 ft.

43. $s^2 = 64h$

$$s^2 = 64\left(\frac{9}{12}\right)$$

$$s^2 = 48$$

$$s = \pm\sqrt{48}$$

$$s = \pm\sqrt{16}\sqrt{3}$$

$$s = \pm 4\sqrt{3}$$

The speed cannot be $-4\sqrt{3}$.
Therefore the speed is
$4\sqrt{3}$ ft/sec.

45. $V = P(1 + r)^t$

    $121 = 100(1 + r)^2$

    $\dfrac{121}{100} = \dfrac{100(1 + r)^2}{100}$

    $1.21 = (1 + r)^2$

    $1 + r = \pm\sqrt{1.21}$

    $1 + r = \pm 1.1$

    $r = -1 \pm 1.1$

    $r = -1 + 1.1$ or $r = -1 - 1.1$

    $r = 0.1$ $r = -2.1$

The rate cannot be -2.1.
Therefore the rate is 0.1, or 10%.

Problem Set 9.2, pp. 416-417

1. $\frac{1}{2}(-8) = -4$ and $(-4)^2 = 16.$

   $x^2 - 8x + 16 = (x - 4)^2$

3. $\frac{1}{2}(12) = 6$ and $6^2 = 36.$

   $m^2 + 12m + 36 = (m + 6)^2$

5. $\frac{1}{2}(-2) = -1$ and $(-1)^2 = 1.$

   $r^2 - 2r + 1 = (r - 1)^2$

7. $\frac{1}{2}(1) = \frac{1}{2}$ and $(\frac{1}{2})^2 = \frac{1}{4}.$

   $y^2 + y + \frac{1}{4} = (y + \frac{1}{2})^2$

9. $\frac{1}{2}(3) = \frac{3}{2}$ and $(\frac{3}{2})^2 = \frac{9}{4}.$

   $p^2 + 3p + \frac{9}{4} = (p + \frac{3}{2})^2$

11. $\frac{1}{2}(-\frac{4}{5}) = -\frac{2}{5}$ and $(-\frac{2}{5})^2 = \frac{4}{25}.$

   $z^2 - \frac{4}{5}z + \frac{4}{25} = (z - \frac{2}{5})^2$

13.   (a) $x^2 - 4x + 3 = 0$

$x^2 - 4x = -3$

$x^2 - 4x + 4 = -3 + 4$

$(x - 2)^2 = 1$

$x - 2 = \pm 1$

$x = 2 \pm 1$

$x = 2 + 1$ or $x = 2 - 1$

$x = 3$ $x = 1$

(b) $x^2 - 4x + 3 = 0$

$(x - 1)(x - 3) = 0$

$x - 1 = 0$ or $x - 3 = 0$

$x = 1$ $x = 3$

15.   (a) $x^2 + 6x + 9 = 0$

$(x + 3)^2 = 0$

$x + 3 = \pm\sqrt{0}$

$x + 3 = 0$

$x = -3$

(b) $x^2 + 6x + 9 = 0$

$(x + 3)(x + 3) = 0$

$x + 3 = 0$ or $x + 3 = 0$

$x = -3$ $x = -3$

17.   (a) $x^2 - 10x = 0$

$x^2 - 10x + 25 = 0 + 25$

$(x - 5)^2 = 25$

$x - 5 = \pm\sqrt{25}$

$x - 5 = \pm 5$

$x = 5 \pm 5$

$x = 5 + 5$ or $x = 5 - 5$

$x = 10$ $x = 0$

(b) $x^2 - 10x = 0$

$x(x - 10) = 0$

$x = 0$ or $x - 10 = 0$

$x = 10$

19. $x^2 - 4x + 2 = 0$

$x^2 - 4x = -2$

$x^2 - 4x + 4 = -2 + 4$

$(x - 2)^2 = 2$

$x - 2 = \pm\sqrt{2}$

$x = 2 \pm \sqrt{2}$

21. $y^2 + 2y - 6 = 0$

$y^2 + 2y = 6$

$y^2 + 2y + 1 = 6 + 1$

$(y + 1)^2 = 7$

$y + 1 = \pm\sqrt{7}$

$y = -1 \pm \sqrt{7}$

23. $m^2 + 10m + 13 = 0$

$m^2 + 10m = -13$

$m^2 + 10m + 25 = -13 + 25$

$(m + 5)^2 = 12$

$m + 5 = \pm\sqrt{12}$

$m + 5 = \pm 2\sqrt{3}$

$m = -5 \pm 2\sqrt{3}$

25. $4x^2 + 8x = 3$

$\dfrac{4x^2}{4} + \dfrac{8x}{4} = \dfrac{3}{4}$

$x^2 + 2x = \dfrac{3}{4}$

$x^2 + 2x + 1 = \dfrac{3}{4} + 1$

$(x + 1)^2 = \dfrac{7}{4}$

$x + 1 = \pm\sqrt{\dfrac{7}{4}}$

$x + 1 = \pm\dfrac{\sqrt{7}}{2}$

$x = -1 \pm \dfrac{\sqrt{7}}{2}$

27. $2p^2 + 4p = 11$

$\dfrac{2p^2}{2} + \dfrac{4p}{2} = \dfrac{11}{2}$

$p^2 + 2p = \dfrac{11}{2}$

$p^2 + 2p + 1 = \dfrac{11}{2} + 1$

$(p + 1)^2 = \dfrac{13}{2}$

29. $3r^2 - 6r - 5 = 0$

$3r^2 - 6r = 5$

$\dfrac{3r^2}{3} - \dfrac{6r}{3} = \dfrac{5}{3}$

$r^2 - 2r = \dfrac{5}{3}$

$r^2 - 2r + 1 = \dfrac{5}{3} + 1$

$$p + 1 = \pm\sqrt{\frac{13}{2}}$$

$$p + 1 = \pm\frac{\sqrt{13}}{\sqrt{2}} \cdot \frac{\sqrt{2}}{\sqrt{2}}$$

$$p + 1 = \pm\frac{\sqrt{26}}{2}$$

$$p = -1 \pm \frac{\sqrt{26}}{2}$$

$$(r - 1)^2 = \frac{8}{3}$$

$$r - 1 = \pm\sqrt{\frac{8}{3}}$$

$$r - 1 = \pm\frac{\sqrt{8}}{\sqrt{3}}$$

$$r - 1 = \pm\frac{2\sqrt{2}}{\sqrt{3}} \cdot \frac{\sqrt{3}}{\sqrt{3}}$$

$$r - 1 = \pm\frac{2\sqrt{6}}{3}$$

$$r = 1 \pm \frac{2\sqrt{6}}{3}$$

31. $k^2 + k - 1 = 0$

$$k^2 + k \quad = 1$$

$$k^2 + k + \frac{1}{4} = 1 + \frac{1}{4}$$

$$\left(k + \frac{1}{2}\right)^2 = \frac{5}{4}$$

$$k + \frac{1}{2} = \pm\sqrt{\frac{5}{4}}$$

$$k + \frac{1}{2} = \pm\frac{\sqrt{5}}{2}$$

$$k = -\frac{1}{2} \pm \frac{\sqrt{5}}{2}$$

33. $2x^2 + 2x - 5 = 0$

$$2x^2 + 2x \quad = 5$$

$$\frac{2x^2}{2} + \frac{2x}{2} \quad = \frac{5}{2}$$

$$x^2 + x \quad = \frac{5}{2}$$

$$x^2 + x + \frac{1}{4} = \frac{5}{2} + \frac{1}{4}$$

$$\left(x + \frac{1}{2}\right)^2 = \frac{11}{4}$$

$$x + \frac{1}{2} = \pm\sqrt{\frac{11}{4}}$$

$$x + \frac{1}{2} = \pm\frac{\sqrt{11}}{2}$$

$$x = -\frac{1}{2} \pm \frac{\sqrt{11}}{2}$$

35.
$$3y^2 = 2y + 1$$
$$3y^2 - 2y = 1$$
$$\frac{3y^2}{3} - \frac{2y}{3} = \frac{1}{3}$$
$$y^2 - \frac{2}{3}y = \frac{1}{3}$$
$$y^2 - \frac{2}{3}y + \frac{1}{9} = \frac{1}{3} + \frac{1}{9}$$
$$\left(y - \frac{1}{3}\right)^2 = \frac{4}{9}$$
$$y - \frac{1}{3} = \pm\sqrt{\frac{4}{9}}$$
$$y - \frac{1}{3} = \pm\frac{2}{3}$$
$$y = \frac{1}{3} \pm \frac{2}{3}$$

$$y = \frac{1}{3} + \frac{2}{3} \quad \text{or} \quad y = \frac{1}{3} - \frac{2}{3}$$
$$y = 1 \qquad\qquad y = -\frac{1}{3}$$

37.
$$m^2 + 6m + 13 = 0$$
$$m^2 + 6m = -13$$
$$m^2 + 6m + 9 = -13 + 9$$
$$(m + 3)^2 = -4$$
$$m + 3 = \pm\sqrt{-4}$$
$$m = -3 \pm \sqrt{-4}$$

Since $\sqrt{-4}$ is not a real number, there is no real solution.

39.
$$4p = -5p^2 + 4$$
$$5p^2 + 4p = 4$$
$$\frac{5p^2}{5} + \frac{4p}{5} = \frac{4}{5}$$
$$p^2 + \frac{4}{5}p = \frac{4}{5}$$
$$p^2 + \frac{4}{5}p + \frac{4}{25} = \frac{4}{5} + \frac{4}{25}$$
$$\left(p + \frac{2}{5}\right)^2 = \frac{24}{25}$$
$$p + \frac{2}{5} = \pm\sqrt{\frac{24}{25}}$$

41.
$$3t^2 - 8t + 6 = 0$$
$$3t^2 - 8t = -6$$
$$\frac{3t^2}{3} - \frac{8t}{3} = \frac{-6}{3}$$
$$t^2 - \frac{8}{3}t = -2$$
$$t^2 - \frac{8}{3}t + \frac{16}{9} = -2 + \frac{16}{9}$$
$$\left(t - \frac{4}{3}\right)^2 = -\frac{2}{9}$$
$$t - \frac{4}{3} = \pm\sqrt{-\frac{2}{9}}$$
$$t = \frac{4}{3} \pm \sqrt{-\frac{2}{9}}$$

$$p + \frac{2}{5} = \pm \frac{\sqrt{24}}{5}$$

$$p + \frac{2}{5} = \pm \frac{2\sqrt{6}}{5}$$

$$p = -\frac{2}{5} \pm \frac{2\sqrt{6}}{5}$$

Since $\sqrt{-\frac{2}{9}}$ is not a real number, there is no real solution.

43. Check $x = 2 + \sqrt{3}$ in $x^2 - 4x + 1 = 0$.

$$(2 + \sqrt{3})^2 - 4(2 + \sqrt{3}) + 1 = 0$$

$$4 + 4\sqrt{3} + 3 - 8 - 4\sqrt{3} + 1 = 0$$

$$0 = 0 \quad \text{True}$$

Check $x = 2 - \sqrt{3}$ in $x^2 - 4x + 1 = 0$.

$$(2 - \sqrt{3})^2 - 4(2 - \sqrt{3}) + 1 = 0$$

$$4 - 4\sqrt{3} + 3 - 8 + 4\sqrt{3} + 1 = 0$$

$$0 = 0 \quad \text{True}$$

45. x = the number

$$x^2 = 2x + 26$$

$$x^2 - 2x = 26$$

$$x^2 - 2x + 1 = 26 + 1$$

$$(x - 1)^2 = 27$$

$$x - 1 = \pm\sqrt{27}$$

$$x - 1 = \pm 3\sqrt{3}$$

$$x = 1 \pm 3\sqrt{3}$$

Since the number must be positive, the solution is $x = 1 + 3\sqrt{3}$.

47. x = first number

$x + 3$ = second number

Product = 7

$$x(x + 3) = 7$$

$$x^2 + 3x = 7$$

$$x^2 + 3x + \frac{9}{4} = 7 + \frac{9}{4}$$

$$\left(x + \frac{3}{2}\right)^2 = \frac{37}{4}$$

$$x + \frac{3}{2} = \pm\sqrt{\frac{37}{4}}$$

$$x + \frac{3}{2} = \pm\frac{\sqrt{37}}{2}$$

$$x = -\frac{3}{2} \pm \frac{\sqrt{37}}{2}$$

Since the numbers must be positive, the first number is $x = -\frac{3}{2} + \frac{\sqrt{37}}{2}$ and the second is

$$x + 3 = -\frac{3}{2} + \frac{\sqrt{37}}{2} + 3 = \frac{3}{2} + \frac{\sqrt{37}}{2} .$$

Problem Set 9.3, pp. 422-424

1. $2x^2 + 9x = -4$

    $2x^2 + 9x + 4 = 0$

    $a = 2, b = 9, c = 4$

3. $x^2 + 4 = 4x$

    $x^2 - 4x + 4 = 0$

    $a = 1, b = -4, c = 4$

5. $x^2 = x$

    $x^2 - x = 0$

    $a = 1, b = -1, c = 0$

7. $9(x^2 + x) - 25 = 9x$

    $9x^2 + 9x - 25 = 9x$

      $9x^2 - 25 = 0$

    $a = 9, b = 0, c = -25$

9. $(2x + 3)^2 = 4(3x + 2)$

    $4x^2 + 12x + 9 = 12x + 8$

        $4x^2 + 1 = 0$

    $a = 4, b = 0, c = 1$

11. $2(x^2 + 1) - x = 4x - 3 + x^2$

    $2x^2 + 2 - x = 4x - 3 + x^2$

    $x^2 - 5x + 5 = 0$

    $a = 1, b = -5, c = 5$

13. (a) $3x^2 + x = 2$

    $3x^2 + x - 2 = 0$

$$x = \frac{-1 \pm \sqrt{1^2 - 4(3)(-2)}}{2(3)}$$

$$x = \frac{-1 \pm \sqrt{1 + 24}}{6}$$

$$x = \frac{-1 \pm \sqrt{25}}{6}$$

(b) $3x^2 + x = 2$

    $3x^2 + x - 2 = 0$

    $(3x - 2)(x + 1) = 0$

    $3x - 2 = 0$ or $x + 1 = 0$

        $3x = 2$ $x = -1$

        $x = \frac{2}{3}$

$$x = \frac{-1 \pm 5}{6}$$

$$x = \frac{-1 + 5}{6} \quad \text{or} \quad x = \frac{-1 - 5}{6}$$

$$x = \frac{2}{3} \qquad\qquad x = -1$$

15. (a) $p^2 - 6p = 0$

$$p = \frac{-(-6) \pm \sqrt{(-6)^2 - 4(1)(0)}}{2(1)}$$

$$p = \frac{6 \pm \sqrt{36 - 0}}{2}$$

$$p = \frac{6 \pm 6}{2}$$

$$p = \frac{6 + 6}{2} \quad \text{or} \quad p = \frac{6 - 6}{2}$$

$$p = 6 \qquad\qquad p = 0$$

(b) $p^2 - 6p = 0$

$$p(p - 6) = 0$$

$$p = 0 \quad \text{or} \quad p - 6 = 0$$

$$p = 6$$

17. (a) $r^2 - 4 = 0$

$$r = \frac{-0 \pm \sqrt{0^2 - 4(1)(-4)}}{2(1)}$$

$$r = \frac{0 \pm \sqrt{0 + 16}}{2}$$

$$r = \frac{\pm 4}{2}$$

$$r = \pm 2$$

(b) $r^2 - 4 = 0$

$$(r - 2)(r + 2) = 0$$

$$r - 2 = 0 \quad \text{or} \quad r + 2 = 0$$

$$r = 2 \qquad\qquad r = -2$$

19. $x^2 + 3x + 1 = 0$

$$x = \frac{-3 \pm \sqrt{3^2 - 4(1)(1)}}{2(1)}$$

$$x = \frac{-3 \pm \sqrt{9 - 4}}{2}$$

$$x = \frac{-3 \pm \sqrt{5}}{2}$$

21. $x^2 = 5x - 1$

$x^2 - 5x + 1 = 0$

$$x = \frac{-(-5) \pm \sqrt{(-5)^2 - 4(1)(1)}}{2(1)}$$

$$x = \frac{5 \pm \sqrt{25 - 4}}{2}$$

$$x = \frac{5 \pm \sqrt{21}}{2}$$

23. $2m^2 - 3m - 1 = 0$

$$m = \frac{-(-3) \pm \sqrt{(-3)^2 - 4(2)(-1)}}{2(2)}$$

$$m = \frac{3 \pm \sqrt{9 + 8}}{4}$$

$$m = \frac{3 \pm \sqrt{17}}{4}$$

25. $3p^2 + p = 1$

$3p^2 + p - 1 = 0$

$$p = \frac{-1 \pm \sqrt{1^2 - 4(3)(-1)}}{2(3)}$$

$$p = \frac{-1 \pm \sqrt{1 + 12}}{6}$$

$$p = \frac{-1 \pm \sqrt{13}}{6}$$

27. $x^2 - 4x + 4 = 0$

$$x = \frac{-(-4) \pm \sqrt{(-4)^2 - 4(1)(4)}}{2(1)}$$

$$x = \frac{4 \pm \sqrt{16 - 16}}{2}$$

$$x = \frac{4 \pm \sqrt{0}}{2}$$

$$x = \frac{4 \pm 0}{2}$$

29. $r^2 - 2r - 7 = 0$

$$r = \frac{-(-2) \pm \sqrt{(-2)^2 - 4(1)(-7)}}{2(1)}$$

$$r = \frac{2 \pm \sqrt{4 + 28}}{2}$$

$$r = \frac{2 \pm \sqrt{32}}{2}$$

$$r = \frac{2 \pm 4\sqrt{2}}{2}$$

$$x = \frac{4}{2}$$

$$x = 2$$

$$r = \frac{2(1 \pm 2\sqrt{2})}{2}$$

$$r = 1 \pm 2\sqrt{2}$$

31. $2t^2 - 4t - 1 = 0$

$$t = \frac{-(-4) \pm \sqrt{(-4)^2 - 4(2)(-1)}}{2(2)}$$

$$t = \frac{4 \pm \sqrt{16 + 8}}{4}$$

$$t = \frac{4 \pm \sqrt{24}}{4}$$

$$t = \frac{4 \pm 2\sqrt{6}}{4}$$

$$t = \frac{2(2 \pm \sqrt{6})}{4}$$

$$t = \frac{2 \pm \sqrt{6}}{2}$$

33. $5y^2 - 10y + 4 = 0$

$$y = \frac{-(-10) \pm \sqrt{(-10)^2 - 4(5)(4)}}{2(5)}$$

$$y = \frac{10 \pm \sqrt{100 - 80}}{10}$$

$$y = \frac{10 \pm \sqrt{20}}{10}$$

$$y = \frac{10 \pm 2\sqrt{5}}{10}$$

$$y = \frac{2(5 \pm \sqrt{5})}{10}$$

$$y = \frac{5 \pm \sqrt{5}}{5}$$

35. $m^2 - 18 = 0$

$$m = \frac{-0 \pm \sqrt{0^2 - 4(1)(-18)}}{2(1)}$$

$$m = \frac{0 \pm \sqrt{0 + 72}}{2}$$

$$m = \frac{\pm\sqrt{72}}{2}$$

$$m = \frac{\pm 6\sqrt{2}}{2}$$

$$m = \pm 3\sqrt{2}$$

37. $4p^2 + 6p + 1 = 0$

$$p = \frac{-6 \pm \sqrt{6^2 - 4(4)(1)}}{2(4)}$$

$$p = \frac{-6 \pm \sqrt{36 - 16}}{8}$$

$$p = \frac{-6 \pm \sqrt{20}}{8}$$

$$p = \frac{-6 \pm 2\sqrt{5}}{8}$$

$$p = \frac{2(-3 \pm \sqrt{5})}{8}$$

$$p = \frac{-3 \pm \sqrt{5}}{4}$$

39.
$$5k = 4 - 6k^2$$

$$6k^2 + 5k - 4 = 0$$

$$k = \frac{-5 \pm \sqrt{5^2 - 4(6)(-4)}}{2(6)}$$

$$k = \frac{-5 \pm \sqrt{25 + 96}}{12}$$

$$k = \frac{-5 \pm \sqrt{121}}{12}$$

$$k = \frac{-5 \pm 11}{12}$$

$$k = \frac{-5 + 11}{12} \quad \text{or} \quad k = \frac{-5 - 11}{12}$$

$$k = \frac{1}{2} \qquad\qquad k = -\frac{4}{3}$$

41.
$$0 = 9z^2 - 6z - 1$$

$$z = \frac{-(-6) \pm \sqrt{(-6)^2 - 4(9)(-1)}}{2(9)}$$

$$z = \frac{6 \pm \sqrt{36 + 36}}{18}$$

$$z = \frac{6 \pm \sqrt{72}}{18}$$

$$z = \frac{6 \pm 6\sqrt{2}}{18}$$

$$z = \frac{6(1 \pm \sqrt{2})}{18}$$

$$z = \frac{1 \pm \sqrt{2}}{3}$$

43.
$$4(x^2 + 2) = 7 - 7x$$

$$4x^2 + 8 = 7 - 7x$$

$$4x^2 + 7x + 1 = 0$$

$$x = \frac{-7 \pm \sqrt{7^2 - 4(4)(1)}}{2(4)}$$

$$x = \frac{-7 \pm \sqrt{49 - 16}}{8}$$

$$x = \frac{-7 \pm \sqrt{33}}{8}$$

45.
$$(y + 4)(y - 3) = -(y + 3)$$

$$y^2 + y - 12 = -y - 3$$

$$y^2 + 2y - 9 = 0$$

$$y = \frac{-2 \pm \sqrt{2^2 - 4(1)(-9)}}{2(1)}$$

$$y = \frac{-2 \pm \sqrt{4 + 36}}{2}$$

$$y = \frac{-2 \pm \sqrt{40}}{2}$$

$$y = \frac{-2 \pm 2\sqrt{10}}{2}$$

$$y = -1 \pm \sqrt{10}$$

47. $\frac{1}{6}x^2 - \frac{1}{3}x - 1 = 0$

$6(\frac{1}{6}x^2 - \frac{1}{3}x - 1) = 6(0)$

$x^2 - 2x - 6 = 0$

$x = \dfrac{-(-2) \pm \sqrt{(-2)^2 - 4(1)(-6)}}{2(1)}$

$x = \dfrac{2 \pm \sqrt{4 + 24}}{2}$

$x = \dfrac{2 \pm \sqrt{28}}{2}$

$x = \dfrac{2 \pm 2\sqrt{7}}{2}$

$x = 1 \pm \sqrt{7}$

49. $\frac{1}{4}r^2 - \frac{1}{3}r - \frac{5}{12} = 0$

$12(\frac{1}{4}r^2 - \frac{1}{3}r - \frac{5}{12}) = 12(0)$

$3r^2 - 4r - 5 = 0$

$r = \dfrac{-(-4) \pm \sqrt{(-4)^2 - 4(3)(-5)}}{2(3)}$

$r = \dfrac{4 \pm \sqrt{16 + 60}}{6}$

$r = \dfrac{4 \pm 2\sqrt{19}}{6}$

$r = \dfrac{2(2 \pm \sqrt{19})}{6}$

$r = \dfrac{2 \pm \sqrt{19}}{3}$

51. $\frac{p^2}{2} + p = \frac{5}{4}$

$4(\frac{p^2}{2} + p) = 4(\frac{5}{4})$

$2p^2 + 4p = 5$

$2p^2 + 4p - 5 = 0$

$p = \dfrac{-4 \pm \sqrt{4^2 - 4(2)(-5)}}{2(2)}$

$p = \dfrac{-4 \pm \sqrt{16 + 40}}{4}$

$p = \dfrac{-4 \pm 2\sqrt{14}}{4}$

53. $6 + \frac{3}{t} = \frac{4}{t^2}$

$t^2(6 + \frac{3}{t}) = t^2(\frac{4}{t^2})$

$6t^2 + 3t = 4$

$6t^2 + 3t - 4 = 0$

$t = \dfrac{-3 \pm \sqrt{3^2 - 4(6)(-4)}}{2(6)}$

$t = \dfrac{-3 \pm \sqrt{9 + 96}}{12}$

$t = \dfrac{-3 \pm \sqrt{105}}{12}$

$$p = \frac{2(-2 \pm \sqrt{14})}{4}$$

$$p = \frac{-2 \pm \sqrt{14}}{2}$$

55. $x^2 + 1 = 0$

$$x = \frac{-0 \pm \sqrt{0^2 - 4(1)(1)}}{2(1)}$$

$$x = \frac{0 \pm \sqrt{0 - 4}}{2}$$

$$x = \frac{\pm \sqrt{-4}}{2}$$

Since $\sqrt{-4}$ is not a real number, the equation has no real solution.

57. $y^2 - 2y + 2 = 0$

$$y = \frac{-(-2) \pm \sqrt{(-2)^2 - 4(1)(2)}}{2(1)}$$

$$y = \frac{2 \pm \sqrt{4 - 8}}{2}$$

$$y = \frac{2 \pm \sqrt{-4}}{2}$$

Since $\sqrt{-4}$ is not a real number, the equation has no real solution.

59. $2m^2 + 3m + 5 = 0$

$$m = \frac{-3 \pm \sqrt{3^2 - 4(2)(5)}}{2(2)} = \frac{-3 \pm \sqrt{9 - 40}}{4} = \frac{-3 \pm \sqrt{-31}}{4}$$

Since $\sqrt{-31}$ is not a real number, the equation has no real solution.

61.   x = first number

2x + 1 = second number

$$x^2 + (2x + 1)^2 = 5$$

$$x^2 + 4x^2 + 4x + 1 = 5$$

$$5x^2 + 4x - 4 = 0$$

$$x = \frac{-4 \pm \sqrt{4^2 - 4(5)(-4)}}{2(5)}$$

$$x = \frac{-4 \pm \sqrt{16 + 80}}{10}$$

$$x = \frac{-4 \pm \sqrt{96}}{10} = \frac{-4 \pm 4\sqrt{6}}{10} = \frac{2(-2 \pm 2\sqrt{6})}{10} = \frac{-2 \pm 2\sqrt{6}}{5}$$

If $x = \frac{-2 + 2\sqrt{6}}{5}$, then $2x + 1 = 2(\frac{-2 + 2\sqrt{6}}{5}) + 1 = \frac{-4 + 4\sqrt{6}}{5} + \frac{5}{5}$

$$= \frac{1 + 4\sqrt{6}}{5}.$$

If $x = \frac{-2 - 2\sqrt{6}}{5}$, then $2x + 1 = 2(\frac{-2 - 2\sqrt{6}}{5}) + 1 = \frac{-4 - 4\sqrt{6}}{5} + \frac{5}{5} = \frac{1 - 4\sqrt{6}}{5}.$

63.   Area = 8

$x(12 - 2x) = 8$

$12x - 2x^2 = 8$

$0 = 2x^2 - 12x + 8$

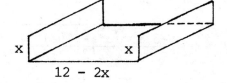

$$x = \frac{-(-12) \pm \sqrt{(-12)^2 - 4(2)(8)}}{2(2)}$$

$$x = \frac{12 \pm \sqrt{144 - 64}}{4}$$

$$x = \frac{12 \pm \sqrt{80}}{4} = \frac{12 \pm 4\sqrt{5}}{4} = \frac{4(3 \pm \sqrt{5})}{4} = 3 \pm \sqrt{5}$$

The height is $x = 3 + \sqrt{5}$ in. or $x = 3 - \sqrt{5}$ in.

65.   x = time fast computer takes

x + 1 = time slow computer takes

Portion of job done by fast computer in 1 hr	+	portion of job done by slow computer in 1 hr	=	portion of job done by both in 1 hr
↓		↓		↓
$\dfrac{1}{x}$	+	$\dfrac{1}{x + 1}$	=	$\dfrac{1}{2}$

$$2x(x + 1)\frac{1}{x} + 2x(x + 1)\frac{1}{x + 1} = 2x(x + 1)\frac{1}{2}$$

$$2(x + 1) + 2x = x(x + 1)$$

$$2x + 2 + 2x = x^2 + x$$

$$0 = x^2 - 3x - 2$$

$$x = \frac{-(-3) \pm \sqrt{(-3)^2 - 4(1)(-2)}}{2(1)} = \frac{3 \pm \sqrt{9 + 8}}{2} = \frac{3 \pm \sqrt{17}}{2}$$

Since the fast computer's time cannot be negative, the fast computer's time is $x = \dfrac{3 + \sqrt{17}}{2}$ hr. The slow computer's time is

$$x + 1 = \frac{3 + \sqrt{17}}{2} + 1 = \frac{3 + \sqrt{17}}{2} + \frac{2}{2} = \frac{5 + \sqrt{17}}{2} \text{ hr.}$$

67. x = speed of the current

	d	r	t	
Upstream	4	6 - x	$\dfrac{4}{6 - x}$	
Downstream	7	6 + x	$\dfrac{7}{6 + x}$	

Since $t = \dfrac{d}{r}$

Total time = 2

$$\frac{4}{6 - x} + \frac{7}{6 + x} = 2$$

$$(6 - x)(6 + x)\frac{4}{6 - x} + (6 - x)(6 + x)\frac{7}{6 + x} = (6 - x)(6 + x)2$$

$$(6 + x)4 + (6 - x)7 = (36 - x^2)2$$

$$24 + 4x + 42 - 7x = 72 - 2x^2$$

$$2x^2 - 3x - 6 = 0$$

$$x = \frac{-(-3) \pm \sqrt{(-3)^2 - 4(2)(-6)}}{2(2)} = \frac{3 \pm \sqrt{9 + 48}}{4} = \frac{3 \pm \sqrt{57}}{4}$$

Since the speed of the current cannot be negative, the speed is $x = \dfrac{3 + \sqrt{57}}{4}$ mph.

69. $h = -16t^2 + 48t$

(a) $16 = -16t^2 + 48t$

$$16t^2 - 48t + 16 = 0$$

$$t^2 - 3t + 1 = 0 \quad \text{Divide by 16}$$

$$t = \frac{-(-3) \pm \sqrt{(-3)^2 - 4(1)(1)}}{2(1)} = \frac{3 \pm \sqrt{9 - 4}}{2} = \frac{3 \pm \sqrt{5}}{2}$$

The height will be 16 ft on the way up at $t = \dfrac{3 - \sqrt{5}}{2}$ sec and on the way down at $t = \dfrac{3 + \sqrt{5}}{2}$ sec.

(b) $64 = -16t^2 + 48t$

$$16t^2 - 48t + 64 = 0$$

$$t^2 - 3t + 4 = 0 \quad \text{Divide by 16}$$

$$t = \frac{-(-3) \pm \sqrt{(-3)^2 - 4(1)(4)}}{2(1)} = \frac{3 \pm \sqrt{9 - 16}}{2} = \frac{3 \pm \sqrt{-7}}{2}$$

Since $\sqrt{-7}$ is not a real number, there is no real solution. The ball never reaches a height of 64 ft.

Problem Set 9.4, pp. 425-426

1. $x^2 = 121$

$x = \pm\sqrt{121}$

$x = \pm 11$

3. $9y^2 = 36y$

$9y^2 - 36y = 0$

$9y(y - 4) = 0$

$9y = 0$ or $y - 4 = 0$

$y = 0$ $y = 4$

5. $z^2 - 8z + 15 = 0$

$(z - 3)(z - 5) = 0$

$z - 3 = 0$ or $z - 5 = 0$

$z = 3$ $z = 5$

7. $x^2 - x = 1$

$x^2 - x - 1 = 0$

$x = \dfrac{-(-1) \pm \sqrt{(-1)^2 - 4(1)(-1)}}{2(1)}$

$x = \dfrac{1 \pm \sqrt{1 + 4}}{2}$

$x = \dfrac{1 \pm \sqrt{5}}{2}$

9. $m^2 = 0$

$m = \pm\sqrt{0}$

$m = \pm 0$

$m = 0$

11. $2y^2 + 9y = 35$

$2y^2 + 9y - 35 = 0$

$(2y - 5)(y + 7) = 0$

$2y - 5 = 0$ or $y + 7 = 0$

$2y = 5$ $y = -7$

$y = \dfrac{5}{2}$

13. $4r^2 - 81 = 0$

  $4r^2 = 81$

  $r^2 = \dfrac{81}{4}$

  $r = \pm\sqrt{\dfrac{81}{4}}$

  $r = \pm\dfrac{9}{2}$

15. $k^2 - 13 = 0$

  $k^2 = 13$

  $k = \pm\sqrt{13}$

17. $6(p + 6)^2 = 54$

  $(p + 6)^2 = 9$

  $p + 6 = \pm\sqrt{9}$

  $p + 6 = \pm3$

  $p = -6 \pm 3$

  $p = -6 + 3$ or $p = -6 - 3$

  $p = -3$ $p = -9$

19. $-t^2 + 32 = 0$

  $-t^2 = -32$

  $t^2 = 32$

  $t = \pm\sqrt{32}$

  $t = \pm4\sqrt{2}$

21. $25m^2 = 40m - 16$

  $25m^2 - 40m + 16 = 0$

  $(5m - 4)^2 = 0$

  $5m - 4 = \pm\sqrt{0}$

  $5m = 4$

  $m = \dfrac{4}{5}$

23. $(6z - 2)^2 = 90$

  $6z - 2 = \pm\sqrt{90}$

  $6z - 2 = \pm3\sqrt{10}$

  $6z = 2 \pm 3\sqrt{10}$

  $z = \dfrac{2 \pm 3\sqrt{10}}{6}$

25. $3x^2 - 7x + 3 = 0$

$$x = \frac{-(-7) \pm \sqrt{(-7)^2 - 4(3)(3)}}{2(3)}$$

$$x = \frac{7 \pm \sqrt{49 - 36}}{6}$$

$$x = \frac{7 \pm \sqrt{13}}{6}$$

27. $(y + 4)^2 = 8y$

$$y^2 + 8y + 16 = 8y$$

$$y^2 + 16 = 0$$

$$y^2 = -16$$

$$y = \pm\sqrt{-16}$$

No real solution

29. $24k^2 + 70k - 75 = 0$

$$k = \frac{-70 \pm \sqrt{70^2 - 4(24)(-75)}}{2(24)}$$

$$k = \frac{-70 \pm \sqrt{4900 + 7200}}{48}$$

$$k = \frac{-70 \pm \sqrt{12,100}}{48}$$

$$k = \frac{-70 \pm 110}{48}$$

$$k = \frac{-70 + 110}{48} \quad \text{or} \quad k = \frac{-70 - 110}{48}$$

$$k = \frac{5}{6} \qquad\qquad k = -\frac{15}{4}$$

31. $(3r + 2)^2 = r^2 + 11r + 4$

$$9r^2 + 12r + 4 = r^2 + 11r + 4$$

$$8r^2 + r = 0$$

$$r(8r + 1) = 0$$

$$r = 0 \quad \text{or} \quad 8r + 1 = 0$$

$$8r = -1$$

$$r = -\frac{1}{8}$$

33. $(2p + 1)(p - 2) = 2p - 6$

$$2p^2 - 3p - 2 = 2p - 6$$

$$2p^2 - 5p + 4 = 0$$

35. $x^2 + \frac{2}{3}x - \frac{1}{2} = 0$

$$6\left(x^2 + \frac{2}{3}x - \frac{1}{2}\right) = 6(0)$$

$$6x^2 + 4x - 3 = 0$$

$$p = \frac{-(-5) \pm \sqrt{(-5)^2 - 4(2)(4)}}{2(2)}$$

$$x = \frac{-4 \pm \sqrt{4^2 - 4(6)(-3)}}{2(6)}$$

$$p = \frac{5 \pm \sqrt{25 - 32}}{4}$$

$$x = \frac{-4 \pm \sqrt{16 + 72}}{12}$$

$$p = \frac{5 \pm \sqrt{-7}}{4}$$

$$x = \frac{-4 \pm \sqrt{88}}{12}$$

No real solution

$$x = \frac{-4 \pm 2\sqrt{22}}{12}$$

$$x = \frac{2(-2 \pm \sqrt{22})}{12}$$

$$x = \frac{-2 \pm \sqrt{22}}{6}$$

Problem Set 9.5, pp. 430-431

1. $\sqrt{-4} = \sqrt{4(-1)} = \sqrt{4}\sqrt{-1} = 2i$

3. $\sqrt{-3} = \sqrt{3(-1)} = \sqrt{3}\sqrt{-1} = \sqrt{3}\,i = i\sqrt{3}$

5. $\sqrt{-20} = \sqrt{4 \cdot 5(-1)} = \sqrt{4}\sqrt{5}\sqrt{-1} = 2\sqrt{5}\,i = 2\,i\sqrt{5}$

7. $-\sqrt{-36} = -\sqrt{36(-1)} = -\sqrt{36}\sqrt{-1} = -6i$

9. $\sqrt{-\frac{1}{9}} = \sqrt{\frac{1}{9}(-1)} = \sqrt{\frac{1}{9}}\sqrt{-1} = \frac{1}{3}i$

11. $\sqrt{-\frac{3}{4}} = \sqrt{\frac{3}{4}(-1)} = \sqrt{\frac{3}{4}}\sqrt{-1} = \frac{\sqrt{3}}{\sqrt{4}}i = \frac{\sqrt{3}}{2}i$

13. True   15. True   17. True

19. $(6 + 2i) + (4 + 5i) = 6 + 2i + 4 + 5i = 10 + 7i$

21. $(-8 + 3i) + (7 - 4i) = -8 + 3i + 7 - 4i = -1 - i$

23. $(-5 - i) + 6i = -5 - i + 6i = -5 + 5i$

25. $(10 + 4i) - (3 + 7i) = 10 + 4i - 3 - 7i = 7 - 3i$

27. $(6 - 5i) - (2 - 5i) = 6 - 5i - 2 + 5i = 4$

29. $15 - (4 + 9i) = 15 - 4 - 9i = 11 - 9i$

31. $2i(6 - 3i) = 2i \cdot 6 - 2i \cdot 3i = 12i - 6i^2 = 12i - 6(-1) = 6 + 12i$

33. $(2 + 7i)(3 + 4i) = 6 + 8i + 21i + 28i^2 = 6 + 29i + 28(-1)$

$$= -22 + 29i$$

35. $(5 + 2i)(6 - i) = 30 - 5i + 12i - 2i^2 = 30 + 7i - 2(-1) = 32 + 7i$

37. $(3 - 4i)(3 + 4i) = 3^2 - (4i)^2 = 9 - 16i^2 = 9 - 16(-1) = 25$

39. $i^3 = i^2 \cdot i = (-1)i = -i$

41. $i^5 = i^2 \cdot i^2 \cdot i = (-1)(-1)i = i$

43. $\dfrac{4}{1 + i} = \dfrac{4}{1 + i} \cdot \dfrac{1 - i}{1 - i} = \dfrac{4 - 4i}{1^2 - i^2} = \dfrac{4 - 4i}{1 - (-1)} = \dfrac{4 - 4i}{2} = \dfrac{4}{2} - \dfrac{4i}{2}$

$$= 2 - 2i$$

45. $\dfrac{9 + 2i}{2 + i} = \dfrac{9 + 2i}{2 + i} \cdot \dfrac{2 - i}{2 - i} = \dfrac{18 - 9i + 4i - 2i^2}{2^2 - i^2} = \dfrac{18 - 5i - 2(-1)}{4 - (-1)}$

$$= \dfrac{20 - 5i}{5}$$

$$= \dfrac{20}{5} - \dfrac{5i}{5}$$

$$= 4 - i$$

47. $\dfrac{4-5i}{3-2i} = \dfrac{4-5i}{3-2i} \cdot \dfrac{3+2i}{3+2i} = \dfrac{12+8i-15i-10i^2}{3^2-(2i)^2} = \dfrac{12-7i-10(-1)}{9-4i^2}$

$$= \dfrac{22-7i}{9-4(-1)}$$

$$= \dfrac{22-7i}{13}$$

$$= \dfrac{22}{13} - \dfrac{7}{13}i$$

49. $\dfrac{3+4i}{5i} = \dfrac{3+4i}{5i} \cdot \dfrac{-5i}{-5i} = \dfrac{-15i-20i^2}{-25i^2} = \dfrac{-15i-20(-1)}{-25(-1)}$

$$= \dfrac{20-15i}{25}$$

$$= \dfrac{20}{25} - \dfrac{15i}{25}$$

$$= \dfrac{4}{5} - \dfrac{3}{5}i$$

51. $x^2 + 25 = 0$

$\quad x^2 = -25$

$\quad x = \pm\sqrt{-25}$

$\quad x = \pm\sqrt{25}\sqrt{-1}$

$\quad x = \pm 5i$

53. $y^2 + 8 = 0$

$\quad y^2 = -8$

$\quad y = \pm\sqrt{-8}$

$\quad y = \pm\sqrt{4}\sqrt{-1}\sqrt{2}$

$\quad y = \pm 2i\sqrt{2}$

55. $(z-4)^2 = -7$

$\quad z - 4 = \pm\sqrt{-7}$

$\quad z - 4 = \pm\sqrt{-1}\sqrt{7}$

$\quad z - 4 = \pm i\sqrt{7}$

$\quad z = 4 \pm i\sqrt{7}$

57. $(3y-8)^2 = -90$

$\quad 3y - 8 = \pm\sqrt{-90}$

$\quad 3y - 8 = \pm\sqrt{9}\sqrt{-1}\sqrt{10}$

$\quad 3y - 8 = \pm 3i\sqrt{10}$

$\quad 3y = 8 \pm 3i\sqrt{10}$

$\quad y = \dfrac{8}{3} \pm \dfrac{3i\sqrt{10}}{3}$

$\quad y = \dfrac{8}{3} \pm i\sqrt{10}$

59. $-2(t + 5)^2 = 72$

$$(t + 5)^2 = -36$$

$$t + 5 = \pm\sqrt{-36}$$

$$t + 5 = \pm 6i$$

$$t = -5 \pm 6i$$

61. $p^2 - 4p + 5 = 0$

$$p = \frac{-(-4) \pm \sqrt{(-4)^2 - 4(1)(5)}}{2(1)}$$

$$p = \frac{4 \pm \sqrt{16 - 20}}{2}$$

$$p = \frac{4 \pm \sqrt{-4}}{2}$$

$$p = \frac{4 \pm 2i}{2}$$

$$p = \frac{4}{2} \pm \frac{2i}{2}$$

$$p = 2 \pm i$$

63. $r^2 + 3r + 4 = 0$

$$r = \frac{-3 \pm \sqrt{3^2 - 4(1)(4)}}{2(1)}$$

$$r = \frac{-3 \pm \sqrt{9 - 16}}{2}$$

$$r = \frac{-3 \pm \sqrt{-7}}{2}$$

$$r = \frac{-3 \pm \sqrt{-1}\sqrt{7}}{2}$$

$$r = \frac{-3 \pm i\sqrt{7}}{2}$$

$$r = -\frac{3}{2} \pm \frac{\sqrt{7}}{2}i$$

65. $2m^2 - 4m + 5 = 0$

$$m = \frac{-(-4) \pm \sqrt{(-4)^2 - 4(2)(5)}}{2(2)}$$

$$m = \frac{4 \pm \sqrt{16 - 40}}{4}$$

$$m = \frac{4 \pm \sqrt{-24}}{4}$$

$$m = \frac{4 \pm \sqrt{4}\sqrt{-1}\sqrt{6}}{4}$$

$$m = \frac{4 \pm 2i\sqrt{6}}{4}$$

$$m = \frac{4}{4} \pm \frac{2i\sqrt{6}}{4}$$

$$m = 1 \pm \frac{\sqrt{6}}{2}i$$

67. $4k^2 + 3 = -2k$

$4k^2 + 2k + 3 = 0$

$$k = \frac{-2 \pm \sqrt{2^2 - 4(4)(3)}}{2(4)} = \frac{-2 \pm \sqrt{4 - 48}}{8} = \frac{-2 \pm \sqrt{-44}}{8}$$

$$k = \frac{-2 \pm \sqrt{4}\sqrt{-1}\sqrt{11}}{8} = \frac{-2 \pm 2i\sqrt{11}}{8} = \frac{-2}{8} \pm \frac{2i\sqrt{11}}{8} = -\frac{1}{4} \pm \frac{\sqrt{11}}{4}i$$

Problem Set 9.6, pp. 438-439

1.

x	$y = x^2 + 1$
3	$y = 3^2 + 1 = 9 + 1 = 10$
2	$y = 2^2 + 1 = 4 + 1 = 5$
1	$y = 1^2 + 1 = 1 + 1 = 2$
0	$y = 0^2 + 1 = 0 + 1 = 1$
-1	$y = (-1)^2 + 1 = 1 + 1 = 2$
-2	$y = (-2)^2 + 1 = 4 + 1 = 5$
-3	$y = (-3)^2 + 1 = 9 + 1 = 10$

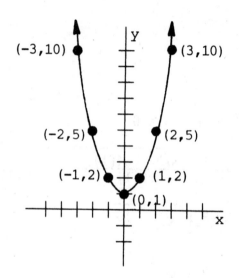

3.

x	$y = -x^2 + 1$
3	$y = -3^2 + 1 = -9 + 1 = -8$
2	$y = -2^2 + 1 = -4 + 1 = -3$
1	$y = -1^2 + 1 = -1 + 1 = 0$
0	$y = -0^2 + 1 = 0 + 1 = 1$
-1	$y = -(-1)^2 + 1 = -1 + 1 = 0$
-2	$y = -(-2)^2 + 1 = -4 + 1 = -3$
-3	$y = -(-3)^2 + 1 = -9 + 1 = -8$

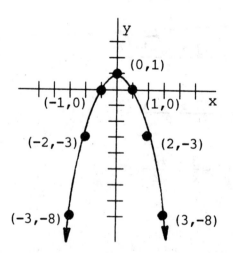

5.

x	$y = 2x^2 - 8$
3	$y = 2(3)^2 - 8 = 2(9) - 8 = 10$
2	$y = 2(2)^2 - 8 = 2(4) - 8 = 0$
1	$y = 2(1)^2 - 8 = 2(1) - 8 = -6$
0	$y = 2(0)^2 - 8 = 2(0) - 8 = -8$
-1	$y = 2(-1)^2 - 8 = 2(1) - 8 = -6$
-2	$y = 2(-2)^2 - 8 = 2(4) - 8 = 0$
-3	$y = 2(-3)^2 - 8 = 2(9) - 8 = 10$

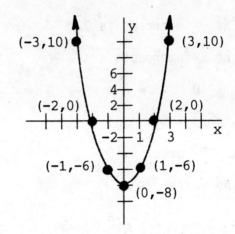

7. $y = x^2 - 4x$

The y-intercept is 0.

The x-intercepts are

$$0 = x^2 - 4x$$

$$0 = x(x - 4)$$

$$x = 0 \quad \text{or} \quad x = 4.$$

The vertex is at

$$x = \frac{-b}{2a} = \frac{-(-4)}{2(1)} = 2,$$

$$y = 2^2 - 4(2) = -4.$$

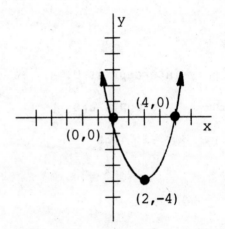

9. $y = x^2 + 6x$

The y-intercept is 0.

The x-intercepts are

$$0 = x^2 + 6x$$

$$0 = x(x + 6)$$

$x = 0$ or $x = -6$.

The vertex is at

$$x = \frac{-b}{2a} = \frac{-6}{2(1)} = -3,$$

$$y = (-3)^2 + 6(-3) = -9.$$

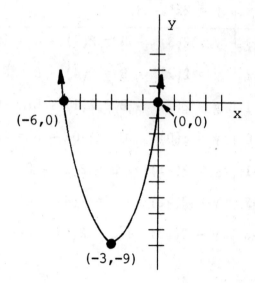

11. $y = -x^2 + 2x$

The y-intercept is 0.

The x-intercepts are

$$0 = -x^2 + 2x$$

$$0 = x^2 - 2x$$

$$0 = x(x - 2)$$

$x = 0$ or $x = 2$.

The vertex is at

$$x = \frac{-b}{2a} = \frac{-2}{2(-1)} = 1,$$

$$y = -1^2 + 2(1) = 1.$$

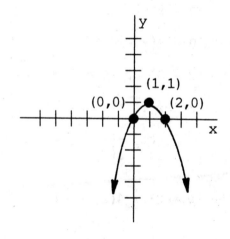

13. $y = x^2 - 2x - 8$

The y-intercept is -8.

The x-intercepts are

$$0 = x^2 - 2x - 8$$

$$0 = (x - 4)(x + 2)$$

$$x = 4 \ \text{ or } \ x = -2.$$

The vertex is at

$$x = \frac{-b}{2a} = \frac{-(-2)}{2(1)} = 1,$$

$$y = 1^2 - 2(1) - 8 = -9.$$

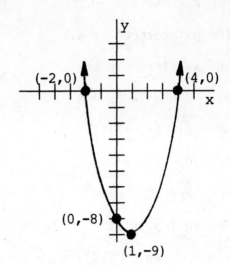

15. $y = -x^2 - 6x - 5$

The y-intercept is -5.

The x-intercepts are

$$0 = -x^2 - 6x - 5$$

$$0 = x^2 + 6x + 5$$

$$0 = (x + 1)(x + 5)$$

$$x = -1 \ \text{ or } \ x = -5.$$

The vertex is at

$$x = \frac{-b}{2a} = \frac{-(-6)}{2(-1)} = -3,$$

$$y = -(-3)^2 - 6(-3) - 5 = 4.$$

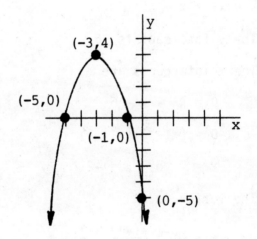

17. $y = x^2 - 4x + 4$

The y-intercept is 4.

The x-intercepts are

$0 = x^2 - 4x + 4$

$0 = (x - 2)^2$

$0 = x - 2$

$2 = x.$

The vertex is at

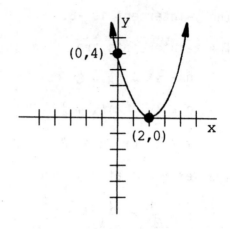

$x = \dfrac{-b}{2a} = \dfrac{-(-4)}{2(1)} = 2,$

$y = 2^2 - 4(2) + 4 = 0.$

19. $y = x^2 - x - 6$

The y-intercept is -6.

The x-intercepts are

$0 = x^2 - x - 6$

$0 = (x - 3)(x + 2)$

$x = 3 \quad \text{or} \quad x = -2.$

The vertex is at

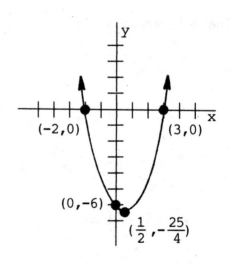

$x = \dfrac{-b}{2a} = \dfrac{-(-1)}{2(1)} = \dfrac{1}{2},$

$y = \left(\dfrac{1}{2}\right)^2 - \left(\dfrac{1}{2}\right) - 6 = -\dfrac{25}{4}.$

21. $y = x^2 - 6x + 10$

The y-intercept is 10.

The x-intercepts are

$$0 = x^2 - 6x + 10$$

$$x = \frac{-(-6) \pm \sqrt{(-6)^2 - 4(1)(10)}}{2(1)}$$

$$x = \frac{6 \pm \sqrt{-4}}{2} .$$

Since $\sqrt{-4}$ is not a real number, there are no x-intercepts.

The vertex is at

$$x = \frac{-b}{2a} = \frac{-(-6)}{2(1)} = 3,$$

$$y = 3^2 - 6(3) + 10 = 1.$$

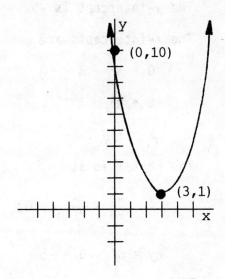

23. $y = 2x^2 + 3x - 2$

The y-intercept is -2.

The x-intercepts are

$$0 = 2x^2 + 3x - 2$$

$$0 = (2x - 1)(x + 2)$$

$$x = \frac{1}{2} \quad \text{or} \quad x = -2.$$

The vertex is at

$$x = \frac{-b}{2a} = \frac{-3}{2(2)} = -\frac{3}{4},$$

$$y = 2(-\frac{3}{4})^2 + 3(-\frac{3}{4}) - 2 = -\frac{25}{8} .$$

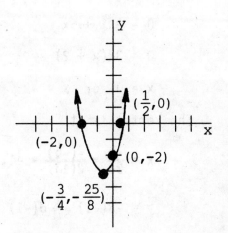

25. $y = x^2 - 5$

The y-intercept is -5.

The x-intercepts are

$$0 = x^2 - 5$$

$$5 = x^2$$

$$x = \pm \sqrt{5}.$$

The vertex is at

$$x = \frac{-b}{2a} = \frac{-0}{2(1)} = 0,$$

$$y = 0^2 - 5 = -5.$$

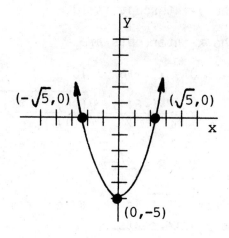

27. $y = -3x^2 - 6x$

The y-intercept is 0.

The x-intercepts are

$$0 = -3x^2 - 6x$$

$$0 = 3x^2 + 6x$$

$$0 = 3x(x + 2)$$

$$x = 0 \quad \text{or} \quad x = -2.$$

The vertex is at

$$x = \frac{-b}{2a} = \frac{-(-6)}{2(-3)} = -1,$$

$$y = -3(-1)^2 - 6(-1) = 3.$$

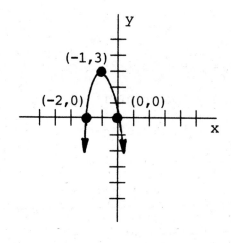

29. $y = \frac{1}{4}x^2$

The y-intercept is 0.

The x-intercepts are

$0 = \frac{1}{4}x^2$

$0 = x^2$

$0 = x.$

The vertex is at

$x = \frac{-b}{2a} = \frac{-0}{2(\frac{1}{4})} = 0,$

$y = \frac{1}{4}(0)^2 = 0.$

x	$y = \frac{1}{4}x^2$
2	$y = \frac{1}{4}(2)^2 = 1$
4	$y = \frac{1}{4}(4)^2 = 4$

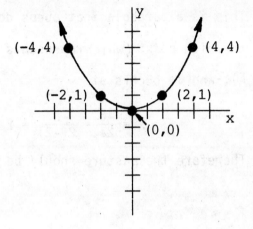

31.

x	$y = x^2$
2	4
1	1
0	0

x	$y = 2x^2$
2	8
1	2
0	0

x	$y = \frac{1}{2}x^2$
2	2
1	$\frac{1}{2}$
0	0

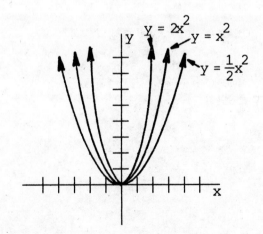

33. A = area of pasture

A = x(8 - 2x)

A = 8x - 2x²

A = -2x² + 8x

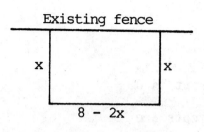

Existing fence

x   x

8 - 2x

This is a parabola that opens downward. The vertex is

the high point (which represents the maximum value of

A), and it occurs at

$$ x = \frac{-b}{2a} = \frac{-8}{2(-2)} = 2. $$

Therefore the pasture should be x = 2 mi by 8 - 2x = 4 mi.